AF556768

Free Radical Biology in Digestive Diseases

Frontiers of Gastrointestinal Research

Vol. 29

Series Editor

Markus M. Lerch Greifswald

Free Radical Biology in Digestive Diseases

Volume Editors

Yuji Naito Kyoto, Japan

Makoto Suematsu Tokyo, Japan

Toshikazu Yoshikawa Kyoto, Japan

36 figures, 3 in color, and 9 tables, 2011

Basel · Freiburg · Paris · London · New York · Bangalore · Bangkok · Shanghai · Singapore · Tokyo · Sydney

Frontiers of Gastrointestinal Research
Founded 1975 by L. van der Reis, San Francisco, Calif.

Dr. Yuji Naito
Molecular Gastroenterology and Hepatology
Kyoto Prefectural University of Medicine
465 Kajii-cho, Kamigyo-ku
Kyoto 602-8566
Japan

Makoto Suematsu
Department of Biochemistry
School of Medicine
Keio University
35 Shinanomachi, Shinjuku-ku
Tokyo 160-8582
Japan

Toshikazu Yoshikawa
Molecular Gastroenterology and Hepatology
Kyoto Prefectural University of Medicine
465 Kajii-cho, Kamigyo-ku
Kyoto 602-8566
Japan

Library of Congress Cataloging-in-Publication Data

Free radical biology in digestive diseases / volume editors, Yuji Naito,
Makoto Suematsu, Toshikazu Yoshikawa.
p. ; cm. -- (Frontiers of gastrointestinal research, ISSN 0302-0665
; v. 29)
Includes bibliographical references and indexes.
ISBN 978-3-8055-9609-1 (hard cover : alk. paper) -- ISBN 978-3-8055-9610-7
(e-ISBN)
1. Digestive organs--Pathophysiology. 2. Free radicals
(Chemistry)--Pathophysiology. 3. Active oxygen--Pathophysiology. I. Naito,
Yuji. II. Suematsu, M. (Makoto), 1957- III. Yoshikawa, Toshikazu, 1947- IV.
Series: Frontiers of gastrointestinal research ; v. 29. 0302-0665
[DNLM: 1. Gastrointestinal Diseases--physiopathology. 2. Oxidative
Stress--physiology. 3. Oxidoreductases--therapeutic use. 4. Reactive
Oxygen Species--adverse effects. W1 FR946E v.29 2011 / WI 140]
RC802.9.F74 2011
616.3--dc22
2010038816

Bibliographic Indices. This publication is listed in bibliographic services, including Current Contents®

www.karger.com
Printed in Switzerland on acid-free and non-aging paper (ISO 9706) by Reinhardt Druck, Basel
ISSN 0302–0665
ISBN 978–3–8055–9609–1
e-ISBN 978–3–8055–9610–7

Contents

Preface

In 1981, Granger et al. [1] demonstrated for the first time that superoxide radicals, oxygen-derived free radicals, are mainly responsible for the increased capillary permeability in the ischemic bowel. In 1985, Itoh and Guth [2] found that SOD, a superoxide dismutase, could significantly protect against gastric mucosal injury induced by a withdrawal for 20 min and reperfusion in rats. In 1989, Yoshikawa et al. [3] have produced a new animal model of reperfusion-induced gastric mucosal injury in rats by applying a vascular clamp to the celiac artery, then removing it. Using this model, they have demonstrated that oxygen-derived free radicals and their scavenging system play a crucial role in the pathogenesis of gastric mucosal injury induced by ischemia-reperfusion. Subsequently, various studies have described the role of free radicals in the development of animal models of digestive diseases. There is a growing body of experimental and clinical data to suggest that the organs of the digestive system may be subjected to considerable oxidative stress associated with acute and chronic inflammation. Although inflammation and ischemia play a key role in producing oxygen-derived free radicals in the digestive organ, the contribution of other factors, such as transition metal imbalance, lipid and glucose metabolic disturbance, and the interaction with gaseous molecules including nitric oxide and carbon monoxide, has also been suggested. Recent studies have demonstrated that several biomarkers indicating oxidative stress-mediated damage, such as 4-hydroxy-2-nonenal, nitrotyrosine, 8-hydroxydeoxyguanosine, and more specific lipid peroxides for free radical reaction, may help in monitoring the degree of disease and planning the design of new therapeutic strategies. In addition, recent advances in 'omics' research (genomics, proteomics, metabolomics, etc.) may bring a breakthrough in the field of gastroenterology and hepatology. Several molecular targets for oxidative stress have been presented by the 'omics'. In this book, we invited several outstanding researchers in the field of free radical biology as well as gastroenterology and hepatology to summarize their work, to review their peers' activity, and to encourage us by their opinions. We thank all the anonymous reviewers for their insightful comments. Finally, we thank Karger Publishers for their cooperation in bringing out this book.

Yuji Naito, Kyoto
Makoto Suematsu, Tokyo
Toshikazu Yoshikawa, Kyoto

References

1 Granger DN, Rutili G, McCord JM: Superoxide radicals in feline intestinal ischemia. Gastroenterology 1981;81:22–29.

2 Itoh M, Guth PH: Role of oxygen-derived free radicals in hemorrhagic shock-induced gastric lesion in rats. Gastroenterology 1985;88:1162–1167.

3 Yoshikawa T, Ueda S, Naito Y, Takahashi S, Oyamada H, Morita Y, Yoneta T, Kondo M: Role of oxygen-derived free radicals in gastric mucosal injury induced by ischemia or ischemia-reperfusion in rats. Free Radic Res Commun 1989;7:285–291.

Naito Y, Suematsu M, Yoshikawa T (eds): Free Radical Biology in Digestive Diseases.
Front Gastrointest Res. Basel, Karger, 2011, vol 29, pp 1–11

Free Radicals and Lipid Peroxidation

Etsuo Niki

Kyoto Prefectural University of Medicine, Kyoto, National Institute of Advanced Industrial Science and Technology, Ikeda, Japan

Abstract

Lipid peroxidation has been implicated in the pathogenesis of various disorders and diseases. Free and ester forms of unsaturated fatty acids and cholesterol are important substrates, and they are oxidized by three distinct mechanisms, namely, free radical-mediated chain reaction, enzyme-mediated oxidation, and nonradical, nonenzymatic oxidation. Each oxidation gives specific products. Furthermore, each oxidation mechanism requires specific antioxidant to inhibit. Lipid peroxidation induces disturbance of membrane fine structure and loss of functions, and it modifies low-density lipoprotein to proatherogenic forms. Lipid peroxidation products are often toxic and modify proteins and DNA bases. Many studies show the elevation of lipid peroxidation products with the progress of diseases, although it has not been elucidated clearly whether lipid peroxidation plays a causal role in the pathogenesis of diseases or it is a consequence of diseases.

There is now ample evidence which shows the involvement of oxidative stress induced by reactive oxygen and nitrogen species, ROS and RNS, respectively, in the pathogenesis of various disorders and diseases. Above all, lipid peroxidation mediated by free radicals has been accepted to play an important role. Among the biological molecules, lipids, especially polyunsaturated fatty acids, are quite susceptible to free radical attack and oxidized by chain mechanism. Cholesterol is also an important substrate of lipid peroxidation. Free and ester forms of unsaturated fatty acids and cholesterol are oxidized by three distinct mechanisms as described below. Lipid peroxidation induces disturbance of fine structure, alteration of integrity, fluidity, and permeability, and functional loss of biomembranes, modifies low-density lipoprotein to proatherogenic and proinflammatory forms, and generates potentially toxic products. Lipid peroxidation products have been shown to be mutagenic and carcinogenic. The secondary reaction products of lipid peroxidation, above all reactive unsaturated carbonyl compounds such as acrolein and 4-hydroxy-2-nonenal (HNE), modify biologically essential molecules such as proteins and DNA bases and enhance the oxidative damage.

At the same time, recent studies revealed that lipid peroxidation products may exert various biological effects such as activation of various transcription factors. In fact, it has been shown that lipid peroxidation products are, like ROS and RNS, double-edged sword and may exert beneficial as well as deleterious effects such as cytotoxic and cytoprotective effects, pro- and anti-atherogenic effects and pro- and anti-inflammatory effects, and pro- and anti-apoptotic effects. However, the physiological significance and role of such effects in vivo have not been unequivocally elucidated and are the subjects of future studies.

Mechanisms and Products of Lipid Peroxidation

Both free and ester forms of fatty acids and cholesterol are oxidized by three distinct mechanisms: they are (1) free radical-mediated oxidation, (2) free radical-independent, nonenzymatic oxidation, and (3) enzymatic oxidation [1]. Specific products are formed from the respective mechanism and specific antioxidants are required to inhibit each type of oxidation.

Free Radical-Mediated Oxidation

The free radical-mediated oxidation proceeds by a chain mechanism, that is, one initiating radical may oxidize many molecules of lipids. Different types of free radicals may initiate the oxidation, but the chain oxidation is propagated by the lipid peroxyl radicals independent of the type of initiating radicals. The mechanism of free radical-mediated lipid oxidation has been studied extensively and is now well understood [1, 2]. The major reactions include (1) abstraction of bisallylic hydrogen to give carbon-centered radical which rearranges rapidly to more stable *cis,trans*-pentadienyl radical, (2) addition of oxygen to the pentadienyl radical to give peroxyl radical, (3) release of oxygen from the peroxyl radical to give oxygen and pentadienyl radical (the adverse reaction 2), which reacts with oxygen to give thermochemically more stable *trans,trans*-peroxyl radical, and (4) intramolecular addition of the peroxyl radical to the double bond to yield cyclic peroxide. The oxidation of linoleates proceeds by a straightforward mechanism to give four conjugated diene hydroperoxides almost quantitatively: 9- and 13-*cis,trans*- and *trans,trans*-hydroperoxyoctadecadienoic acid (HPODE) (fig. 1).

Compared with linoleates, the oxidation of more highly unsaturated fatty acids such as arachidonic acid (20:4), eicosapentaenoic acid (EPA 20:5), and docosahexaenoic acid (DHA 22:6) having more than two double bonds proceeds by more complicated mechanisms to give versatile products including cyclic peroxides and their secondary reaction products containing mixtures of many isomers.

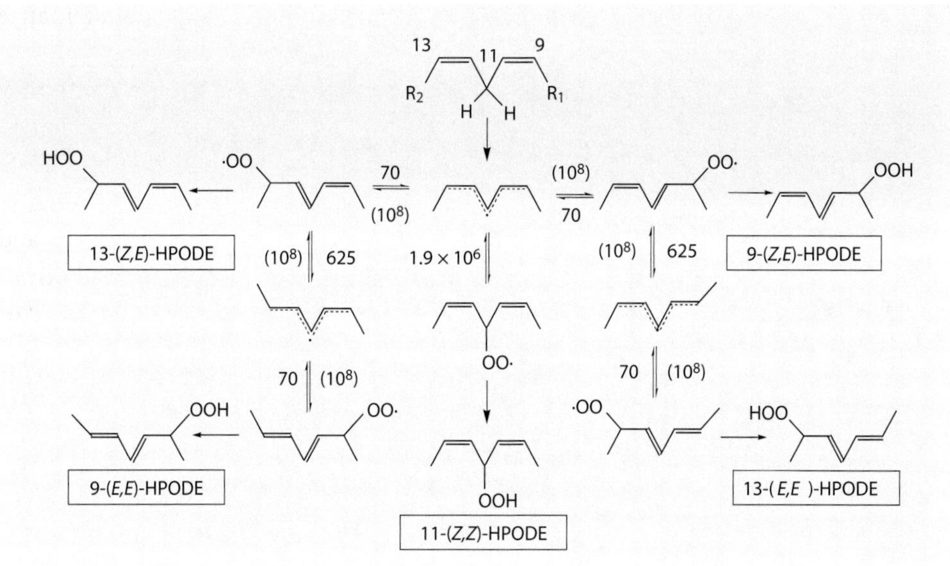

Fig. 1. Reaction pathways in the free radical-mediated oxidation of linoleates.

Nonradical, Nonenzymatic Oxidation

Myeloperoxidase, a heme protein secreted by activated phagocytes, reacts with hydrogen peroxide in the presence of chloride to give hypochlorous acid, which is a strong oxidant, and oxidizes biological molecules by several mechanisms including both free radical and nonradical mechanisms, as shown in figure 2 [3].

Singlet oxygen and ozone oxidize lipids by a nonradical mechanism [4]. Singlet oxygen may be used in photodynamic therapy or cause deleterious damage to the skin. It oxidizes unsaturated compounds by ene-reaction to give hydroperoxides with concomitant double bond migration, with minor side reactions such as 1,4-addition to give 1,4-endoperoxide and 1,2 addition to give dioxetane, which readily decomposes to yield carbonyl compounds. The oxidation of linoleates by singlet oxygen gives 9, 10, 12, and 13-HPODE as major products, 10- and 12-HPODE being specific products of singlet oxygen oxidation [1]. Ozone oxidizes unsaturated compounds to give ozonide and cleavage products.

Enzymatic Oxidation

The enzymatic oxidation of unsaturated lipids by lipoxygenase (LOX), cyclooxygenase (COX), and cytochrome P450 (CYP) is another important type of lipid oxidation. Above all, the oxidation of arachidonic acid by LOX and COX gives biologically

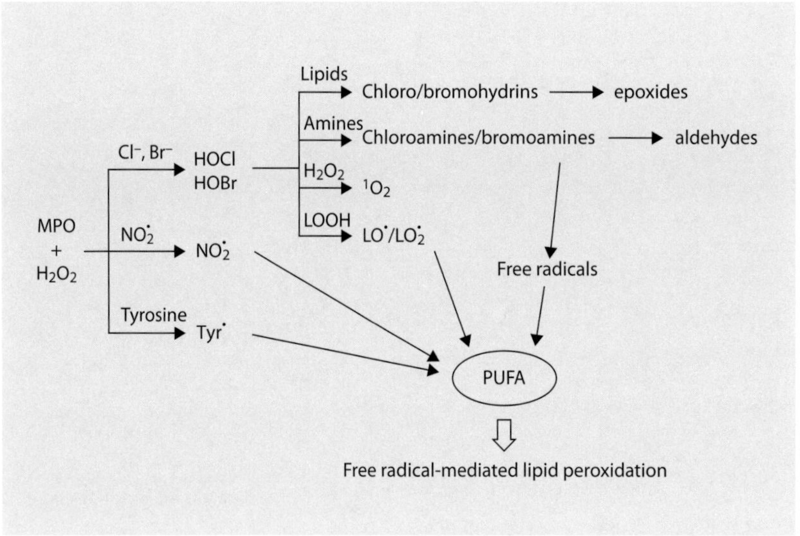

Fig. 2. Main pathways of myeloperoxidase (MPO)-mediated lipid peroxidation.

important various products such as hydroperoxyeicosatetraenoic acid (HPETE), prostaglandins, prostacyclins, thromboxanes, and leukotrienes. The major LOXs are 5-, 12-, and 15-LOX and oxidize arachidonates to give corresponding HPETEs. CYPs induce allylic oxidation and ω-hydroxylation to give 5-, 8-, 9-, 11-, 12-, 15-, 19-, and 20-HETE. HETEs produced by different pathways are summarized in table 1. CYP also mediates epoxidation to produce 5,6-, 8,9-, 11,12-, and 14,15-epoxy-eicosatetraenoic acid from arachidonic acid. The characteristic of enzymatic oxidation is that it proceeds by regio-, stereo-, and enantio-specific mechanisms [5].

Oxidation of Cholesterol

Cholesterol is also oxidized by three mechanisms described above to give various products called oxysterols (fig. 3) [6]. The free radical-mediated oxidation of cholesterol gives 7-hydroperoxycholesterol, 7-hydroxycholesterol, 7-ketocholesterol, cholesterol-5,6-epoxide, and cholestane-3β,5α,6β-triol. Both α- and β-forms of 7-hydroperoxy-, 7-hydroxy-, and 5,6-epoxy-cholesterol are formed. On the other hand, singlet oxygen gives 5-hydroperoxycholesterol. The oxidation of cholesterol by hypochlorous acid gives chlorohydrins and 5,6-dichlorocholesterol as major product [7].

The CYP hydroxylase enzyme oxidizes cholesterol to hydroxycholesterols. Interestingly, 24-hydroxylase, CYP46A1, is present exclusively in neuronal cells in the brain and retina and gives 24(S)-hydroxycholesterol. 25-Hydroxycholesterol is formed by cholesterol 25-hydroxylase, which is not a CYP enzyme. The interconversion

Table 1. HETE formed from arachidonates by various oxidants

	Regio-isomer	Stereo	Enantio
Free radicals	5, 8, 9, 11, 12, 15	*ZE*, ***EE***	*R* = *S*
LOX	5, 12, 15	*ZE*	*R* or *S*
CYP	5, 8, 9, 11, 12, 15, **19**, **20**	*ZE*	*R* or *S*
Singlet oxygen	5, **6**, 8, 9, 11, 12, **14**, 15	*ZE*	

Numbers in regio isomers show the carbon number of eicosatetraenoic acid to which hydroxy group is attached.

between oxysterols such as 7β-cholesterol and 7-ketocholesterol has been found to take place in vivo [8]. These oxysterols have biological functions such as regulating the synthesis and excretion of cholesterol.

Secondary Products of Lipid Peroxidation

The primary products of free radical-mediated lipid peroxidation and LOX-catalyzed oxidation are the hydroperoxides, which are reduced by glutathione peroxidases to the corresponding hydroxides. Accordingly, hydroxides such as HODE and HETE are the most abundant lipid peroxidation products that can be detected from biological fluids. Further, hydroperoxides may undergo secondary reactions to give various products. Some of the important products are discussed below.

Isoprostanes

The group of Morrow and Roberts discovered in 1990 that a series of prostaglandin-like compounds named isoprostanes were formed by the free radical-mediated oxidation of arachidonic acid independent of COX [9]. Isoprostanes are accepted to be the most reliable biomarker of lipid peroxidation in vivo [10]. It has been shown that isoprostanes exert biological activities and may act as mediators of oxidative injury [11]. Several classes of isoprostanes termed D2-, E2-, F2-, J2-, and A2-isoprostanes are formed from H2-isoprostane, of which F2-isoprostanes containing F-type prostane ring have been studied most widely. The 8-, 9-, 11-, and 12-peroxyl radicals derived from arachidonates undergo two consecutive intramolecular cyclizations, oxygen addition, and hydrogen abstraction to form PGG2-like compounds, which are reduced to give F2-isoprostanes. Compounds are denoted as 5-, 8-, 12-, and 15-series regional isomers depending on the carbon atom to which the side chain

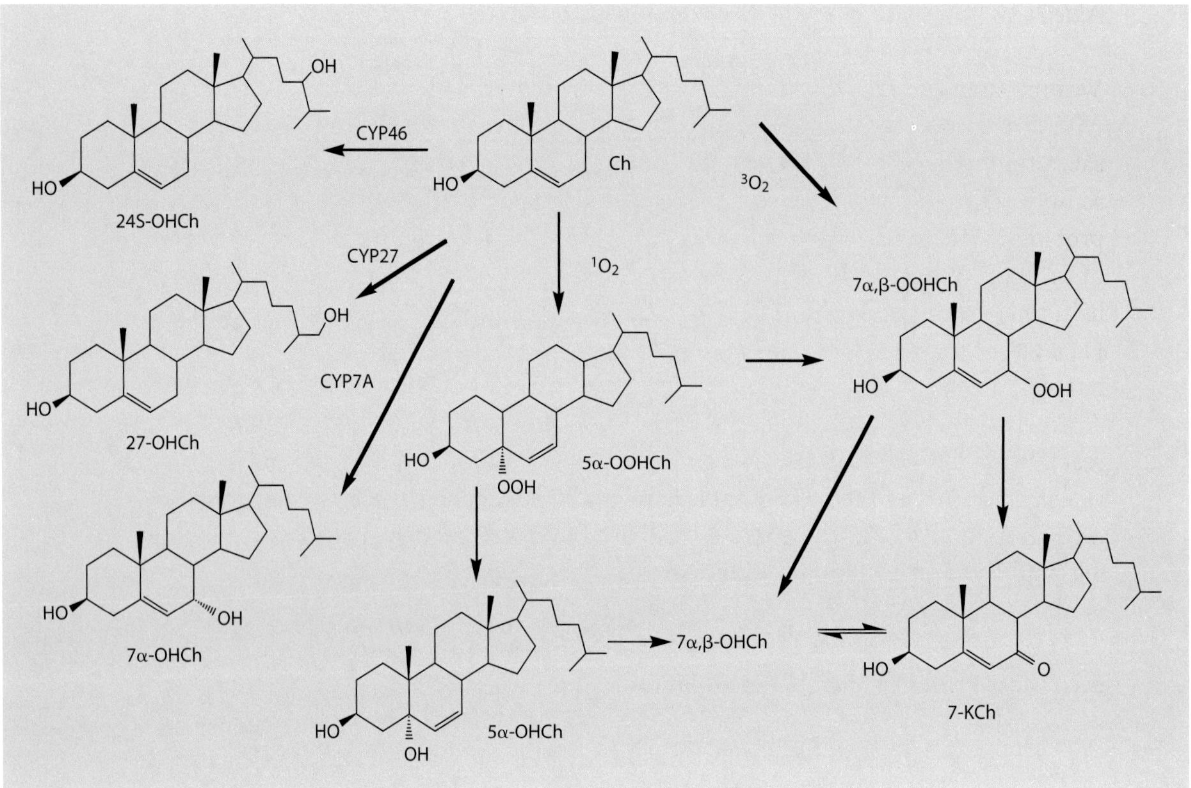

Fig. 3. Oxidation of cholesterol.

hydroxyl is attached. A total of 64 F2-isoprostane isomers can be formed from arachidonates. Isoprostanes are formed primarily in esterified arachidonates and subsequently released by phospholipases, whereas prostaglandins are formed only from free arachidonic acid.

The compounds named F3-isoprostanes and F4-neuroprostanes have similar structures and are generated from EPA and DHA, respectively. EPA and DHA are abundant in neuronal cells and may serve as good biomarker for oxidative damage in the brain.

15-Deoxy-Delta-12,14-Prostaglandin J2

15-Deoxy-delta-12,14-prostaglandin J2 (15d-PGJ2) is another product from arachidonates which has attracted much attention [12]. 15-dPGJ2 is formed by a non-enzymatic, two-step dehydration of PGD2 and can exert versatile biological effects.

Aldehydes

Various small-molecular-weight aldehydes such as acrolein, malondialdehyde, and HNE are formed as secondary products of lipid peroxidation [13]. More than 20 saturated and unsaturated aldehydes have been identified and found in biological samples. The α,β-unsaturated aldehydes are highly reactive and readily react with proteins, DNA, and phospholipids. The modification of amino and thiol groups of proteins and peptides by these aldehydes occurs mainly at cysteine, lysine, and histidine to produce stable covalent-bonded adducts by Michael addition reaction [14]. Carbonyl group may alternatively react with the amino groups to form Schiff bases.

Since its discovery in 1964 [15], HNE has been studied extensively [16]. HNE reacts with thiol and amino groups, and HNE-protein adducts have been detected in mammalian organisms by mass spectrometry-based technique and polyclonal and monoclonal antibodies directed against protein-bound cysteine, lysine or histidine adducts. These adducts have been implicated in the genesis and/or progression of a variety of human disease processes.

It should be added that aldehydes are generated also by another pathway in a myeloperoxidase – hydrogen peroxide – chloride ion system [17].

Nitro-Fatty Acids

Nitro-fatty acids have also attracted attention as one of the lipid oxidation products which react with various biological molecules and elicit multiple biological effects [18]. Two allylic nitro derivatives of linoleic acid were found in human erythrocytes and plasma as both free and ester forms. They were identified as 10 nitro-9-*cis*,12-*cis*-octadecadienoic acid and 12-nitro-9-*cis*,12-*cis*-octadecadieoic acid. 9-Nitro-oleic acid and 10-nitro-oleic acid were also detected, but their plasma concentrations reported by two groups are quite different, approaching 1,000 nM of combined free and esterified 9- and 10-nitro-oleic acid [19] and 1 nM each of free 9- and 10-nitro-oleic acid [20].

Lipid Peroxidation Products as Biomarker for Oxidative Stress in vivo

The levels of lipid peroxidation products in biological fluids and tissues of human subjects have been measured extensively [1]. Previously, thiobarbituric acid-reactive substances, ethane and pentane in breath gas, lipid hydroperoxides, and aldehydes such as malonaldehyde have been often measured as a marker of lipid peroxidation in vivo. Recent advances of mass spectrometry enabled precise and accurate identification and quantification of lipid peroxidation products in their intact forms.

Table 2. Biomarkers of oxidative stress

Lipid	Protein	DNA
Ethane and pentane	Protein carbonyl	Comet assay
in exhaled gas	Hydroperoxide	Thymine glycol
TBARS	Nitro-, chloro-, bromo-amino acid	5-Hydroxyuracil
Conjugated diene	Disulfide -SS-	2-, 8-Hydroxyadenine
Hydroperoxide	-SOH, -SOOH, -SOOOH	8-Hydroxyguanine
Aldehydes	Aldehyde-modified protein	8-Nitro-, chloro-, bromo-guanine
Ketone	Hydroperoxide-modified protein	
Isoprostane	Crosslinked protein	
Neuroprostane	Dityrosine	
Isofuran	Albumin dimer	
Neurofuran	Advanced oxidation products	
HODE	Creatol	
HETE	Myeloperoxidase	
Lyso PC	Lipofuscin	
Oxidized low-density lipoprotein	Cleavage products	
Oxysterols		

TBARS = Thiobarbituric acid-reactive substances.

Biomarkers used for assessment of oxidative damage in vivo are summarized in table 2.

Various lipid peroxidation products have been applied for assessment of lipid peroxidation and oxidative stress status in vivo. In addition to these products, proteins and DNA bases modified by the lipid peroxidation products have also been used as biomarkers. Above all, the adducts of lysine residues and unsaturated aldehydes such as HNE and acrolein have been used most frequently [21]. The adduct of lysine with lipid hydroperoxide has also been identified and found in human atherosclerotic lesions [22]. It was reported that the urinary metabolite of 4-HNE with GSH,

1,4-dihydroxynonene mercapturic acid, might be used as a biomarker of lipid peroxidation [23].

Many studies have explored the correlation between the level of lipid peroxidation products and disease state and their application as biomarkers for early detection and prognosis of diseases [24]. Above all, isoprostanes have been most extensively studied by many groups. In some cases, but not always, a strong correlation was observed. Lipid peroxidation products have advantages and disadvantages as biomarkers. Compared with proteins, a high specificity may not be expected for lipid peroxidation products. In other words, the increase in lipid peroxidation product levels suggests the elevation of oxidative stress status, but it is difficult to specify disease or tissue. In contrast, lipid peroxidation products may be useful as an indicator of overall oxidative stress status and healthy state.

Arachidonic acid is more readily oxidized than linoleic acid, and its oxidation mechanisms are well understood. Isoprostanes are a unique product from the free radical-mediated oxidation of arachidonic acid. One disadvantage of isoprostanes may be that the oxidation of arachidonic acid proceeds by many competing reactions to give numerous products and that isoprostanes are composed of 64 isomers, which make the selectivity of formation and concentration of isoprostane isomers very low. The amount of 8-isoprostane F2α is one or two orders less than HODE or HETE. On the other hand, linoleic acid is the most abundant polyunsaturated fatty acid in vivo and it is oxidized by a straightforward mechanism to give only four isomeric HPODE as primary product almost quantitatively. Thus, the level of HODE after reduction and saponification is higher than any other lipid peroxidation product. Another advantage of HODE is that the antioxidant capacity of the environment can be assessed from the ratio of *cis,trans*-HODE/*trans,trans*-HODE. The higher the antioxidant capacity, the larger this ratio. Therefore, the antioxidant capacity of pure compounds and also complex mixtures in foods, beverages, and supplements in vivo may be assessed from this ratio [25].

In general, higher levels of lipid peroxidation products are observed in patients with various diseases than healthy subjects. Similarly, high lipid peroxidation product levels have been observed in neurological disease patients [26]. These results do not indicate that lipid peroxidation is a cause of diseases, but a positive correlation implies that these levels may be used as a biomarker, at least as a surrogate biomarker for the effect of medical treatment and the efficacy of drugs.

Ideally, the biomarkers and their assay should be simple, specific, accurate, sensitive, reproducible, noninvasive, high-throughput, and inexpensive. There are no such biomarkers that satisfy all of these factors yet. Various specimens have been used including plasma, serum, erythrocyte, lymphocyte, lipoprotein, tissues, cells, cerebrospinal fluid, urine, saliva, sperm, tear, breath gas, and various lavages. Noninvasive samples are recommended, but it is difficult to show where the phenomenon occurs. The standardization of specimens, such as collection, storage, processing, and assay conditions is also important.

It may also be noted that lipid peroxidation products can be absorbed from the diet. Foods contain more or less oxidized lipids, and the measurements of lipid peroxidation products in the biological fluids may be confounded by the diet. It has been reported that dietary hydroxyl fatty acids labeled with 13C are absorbed and found in plasma [26]. It is not clear how other lipid peroxidation products such as isoprostanes and oxysterols are absorbed, circulated, and excreted in urine. It is important to collect samples from the fasted subjects. The potential for artifacts produced during sample collection, processing, storage, and instrumental analyses should be always considered.

Overall, many data show the positive association of biomarkers of lipid peroxidation with oxidative stress status and related various diseases, although inconsistencies between various studies exist. The association of isoprostanes with various diseases has been summarized by Yin [10]. At the same time, it may be noteworthy that recently it has been found that low levels of oxidative stress and lipid peroxidation products may act as good stress by inducing adaptive response and elevating defense capacity through upregulation of glutathione and phase II antioxidant enzymes [27]. The biological effects of lipid peroxidation seem quite complicated. Recent studies reveal that lipid peroxidation products may be cytotoxic and cytoprotective, and proinflammatory and anti-inflammatory. It should also be pointed out that some lipid peroxidation may be controlled, but it is difficult to control and program the time, site, and amount of lipid peroxidation mediated by free radicals. It is hoped that future studies will elucidate the biological effects of lipid peroxidation and how lipid peroxidation products may be used as biomarkers for diagnosis, prognosis, and assessment of efficacy of drugs, foods, and supplements.

References

1 Niki E: Lipid peroxidation: physiological levels and dual biological effects. Free Radic Biol Med 2009;47: 469–484.

2 Yin H, Porter NA: New insights regarding the autoxidation of polyunsaturated fatty acids. Antioxid Redox Signal 2005;7:170–184.

3 Malle E, Marsche G, Arnold J, Davies M: Modification of low-density lipoprotein by myeloperoxidase-derived oxidants and reagent hypochlorous acid. Biochim Biophys Acta 2006; 1761:392–415.

4 Pryor WA, Houk KN, Foote CS, Fukuto JM, Ignarro LJ, Squadrito GL, Davies KJA: Free radical biology and medicine: it's a gas, man! Am J Physiol Integr Comp Physiol 2006;291:R491–R511.

5 Schneider C, Pratt DA, Porter NA, Brash AR: Control of oxygenation in lipoxygenase and cyclooxygenase catalysis. Chem Biol 2007;14:473–488.

6 Murphy RC, Johnson KM: Cholesterol, reactive oxygen species, and the formation of biologically active metabolites. J Biol Chem 2008;283:15521–15525.

7 Heinecke JW, Li W, Mueller DM, Bohrer A, Turk J: Cholesterol chlorohydrin synthesis by the myeloperoxidase-hydrogen peroxide-chloride system: potential markers of lipoprotein oxidatively damaged by phagocytes. Biochemistry. 1994;33: 10127–10136.

8 Larsson H, Bottiger Y, Iuliano L, Diczfalusy U: In vivo interconversion of 7β-cholesterol and 7-ketocholesterol, potential surrogate markers for oxidative stress. Free Radic Biol Med 2007;43:695–701.

9 Morrow JD, Hill KE, Burk RF, Nammour TM, Badr KF, Roberts LJ 2nd: A series of prostaglandin F2-like compounds are produced in vivo in humans by a non-cyclooxygenase, free radical-catalyzed mechanism. Proc Natl Acad Sci USA 1990;87:9383–9387.
10 Yin H: New techniques to detect oxidative stress markers: mass spectrometry-based methods to detect isoprostanes as the gold standard for oxidative stress in vivo. Biofactors 2008;34:109–124.
11 Comporti M, Signorini C Arezzini B, Vecchio D, Monaco B, Gardi C: F2-isoprostanes are not just markers of oxidative stress. Free Radic Biol Med 2008;44:247–256.
12 Uchida K, Shibata T: 15-Deoxy-delta(12,14)-prosta glandin J2: an electrophilic trigger of cellular responses. Chem Res Toxicol 2008;21:138–144.
13 Esterbauer H, Schaur JS, Zollner H: Chemistry and biochemistry of 4-hydroxynonenal, malonaldehyde and related aldehydes. Free Radic Biol Med 1991;11: 81–128.
14 Liebler DC: Protein damage by reactive electrophiles: targets and consequences. Chem Res Toxicol 2008;21:117–128.
15 Schauenstein E, Esterbauer H, Jaag G, Taufer M: The effect of aldehydes on normal and malignant cells. 1st report: hydroxyoctenal, a new fat aldehyde. Chem Mon 1964;95:180–183.
16 Poli G, Schaur RJ, Siems WG, Leonarduzzi G: 4-Hydroxynonenal: a membrane lipid oxidation product of medicinal interest. Medic Res Rev 2008;28:569–631.
17 Hazen SL, Hsu FF, d'Avignon A, Heinecke JW: Human neutrophils employ myeloperoxidase to convert amino acids to a battery of reactive aldehydes: a pathway for aldehyde generation at sites of inflammation. Biochemistry 1998;37:6864–6873.
18 Freeman BA, Baker PR, Schopfer FJ, Woodcock SR, Napolitano A, d'Ischia M: Nitro-fatty acid formation and signaling. J Biol Chem 2008;283:15515–15519.
19 Baker PS, Lin Y, Schopfer FJ, Woodcock SR, Groeger AL, Batthyany C, Sweeney S, Long MH, Iles KE, Baker LMS, Branchaud BP, Chen YE, Freeman BA: Fatty acid transduction of nitric oxide signaling. J Biol Chem 2005;51:42464–42475.
20 Tsikas D, Zoerner A, Mitschke A, Homsi Y, Gutzki FM, Jordan J: Specific GC-MS/MS stable isotope dilution methodology for free 9- and 10-nitro-oleic acid in human plasma challenges previous LC-MS/MS reports. J Chromat B 2009;877:2895–2908.
21 Uchida K: Histidine and lysine as targets of oxidative modification. Amino Acids 2003;25:249–257.
22 Kawai Y, Fujii H, Kato Y, Kodama M, Naito M, Uchida K, Osawa T: Esterifed lipid hydroperoxide-derived modification of protein: formation of a carboxyalkylamide-type lysine adduct in human atherosclerotic lesions. Biochem Biophys Res Commun 2004;313:271–276.
23 Peiro G, Alary A, Cravedi JP, Tathahao E, Steghens JP, Gueraud F: Dihydroxynonene mercapturic acid, a urinary metabolite of 4-hydroxynonenal, as a biomarker of lipid peroxidation. Biofactors 2005;24: 89–96.
24 Dalle-Donne I, Rossi R, Colombo R, Giustarini D, Milzani A: Biomarkers of oxidative damage in human disease. Clin Chem 2006;52:601–623.
25 Yoshida Y, Hayakawa M, Habuchi Y, Itoh N, Niki E: Evaluation of lipophilic antioxidant efficacy in vivo by the biomarkers hydroxyoctadecadienoic acid and isoprostane. Lipids 2007;42:463–472.
26 Montine TJ, Quinn J, Kaye J, Morrow JD: F(2)-isoprostanes as biomarkers of late-onset Alzheimer's disease. J Mol Neurosci 2007;33:114–119.
27 Chen ZH, Yoshida Y, Saito Y, Sekine A, Noguchi N, Niki E: Induction of adaptive response and enhancement of PC12 cell tolerance by 7-hydroxycholesterol and 15-deoxy-delta812,14)-prostaglandin J2 through up-regulation of cellular glutathione via different mechanisms. J Biol Chem 2006;281: 14440–14445.

Etsuo Niki
7-17-7 Kitakarasuyama
Setagaya, Tokyo 157-0061 (Japan)
Tel./Fax +81 3 5313 2555, E-Mail etsuo-niki@aist.go.jp

Naito Y, Suematsu M, Yoshikawa T (eds): Free Radical Biology in Digestive Diseases.
Front Gastrointest Res. Basel, Karger, 2011, vol 29, pp 12–22

Mitochondria as Source of Free Radicals

Hideyuki J. Majima[a,b] · Hiroko P. Indo[a] · Shigeaki Suenaga[a] · Tsuyoshi Kaneko[c] · Hirofumi Matsui[c] · Hsiu Chuan Yen[d] · Toshihiko Ozawa[e]

Departments of [a]Oncology and [b]Space Environmental Medicine, Kagoshima University Graduate School of Medical and Dental Sciences, Kagoshima, [c]Division of Gastroenterology, Graduate School of Comprehensive Human Sciences, University of Tsukuba, Tsukuba, Japan; [d]Department of Medical Biotechnology and Laboratory Science, Chang Gung University, Kwei-shan, Tao-Yuan, Taiwan, ROC; [e]Department of Health Pharmacy, Yokohama College of Pharmacy, Kanagawa, Japan

Abstract

Mitochondria have been considered to be mediators of cell metabolism involved in important life processes, such as aging, cell death, and persistent chronic diseases, in addition to their main function of producing ATP. Chronic diseases cause mutations or deletions in the mitochondrial genome (mt genome), which is believed to accumulate damage from long-term oxidative stress. Because mitochondrial DNA (mtDNA) encodes 13 genes for proteins that comprise a part of the electron transport chain (ETC), damage to these genes causes ETC impairment. Treatment with ETC inhibitors (rotenone, 3-nitropropionic acid, thenoyltrifluoroacetone, antimycin A, and sodium cyanide) generates increased intracellular reactive oxygen species (ROS). A significant increase in ROS was also observed in cells lacking mtDNA and mtDNA-deleted cells compared with their parental cells, although no increase was observed in cybrids. Furthermore, cells transfected with cDNA encoding manganese superoxide dismutase had decreased levels of ROS. These results suggest that more ROS are generated from mitochondria in cells that have the ETC impaired either by inhibition or damage to the mtDNA.

The electron transport chain (ETC) is the complex that produces ATP by oxidative phosphorylation. During electron transport, 2–3% of electrons leak from the ETC, and superoxide is generated when they bind with molecular oxygen. Superoxide is then converted to various other reactive oxygen species (ROS) in chain reactions. Among the ROS, peroxinitrite, which is made by the binding of superoxide and nitric oxide, should play a major role in generating oxidative stress within cells. We proved that the primary intracellular ROS are produced from mitochondria and that lipid

Table 1. List of diseases in which free radicals are implicated

Cancer
Alcoholic liver disease
Crohn's disease
Rheumatoid arthritis
Diabetes
Muscular dystrophy
Cystic fibrosis
Septic shock
Premature babies
Atherosclerosis
Infertility
Cataract
Aging
Hepatitis
ARDS
Ischemia
Neuronal degeneration

peroxidation also occurred at the same location. Rheumatoid arthritis, hepatitis, enteritis, cancer, aging, many neurologic diseases, and chronically persisting digestive diseases have recently been found to be related to oxidative stress. It is becoming evident that these are associated with the generation of ROS from the mitochondrial ETC (partially as a result of mitochondrial DNA, mtDNA, damage), subsequent lipid peroxidation, and cell death.

Recent findings suggest that some refractory diseases of unknown etiology are caused by ROS. These diseases take a chronic rather than an acute course, and show complicated pathologies accompanied by changes in the intracellular oxidation/reduction (redox) status, antioxidative response, signal transduction system, and DNA transcription system, which lead to cell death. A number of studies have suggested the involvement of ROS in the development of many diseases and conditions, such as rheumatoid arthritis, hepatitis, enteritis, cancer, and aging [1, 2]. Many neurologic disorders have also been shown to be related to oxidative stress [1]. Table 1 shows a partial list of oxidative stress-related diseases. Neurologic diseases, such as Alzheimer's disease [3], Parkinson's disease [4–6], and amyotrophic lateral sclerosis (ALS) [7] have recently been found to be associated with damage to the mtDNA or an increase in intracellular ROS. Exposure of organisms to oxidative stress leads to cellular damage and development of certain types of diseases. Does the antioxidative system function as an intracellular defense system? Biologic defense systems can be both enzymatic and nonenzymatic, and scavenging enzymes can detoxify ROS. These systems work together to combat active ROS.

Superoxide Theory

This theory [8] proposes that superoxide ($O_2^{\bullet -}$) is the most important ROS and is based on the discovery of superoxide dismutase, a superoxide-scavenging enzyme, by Fridovich et al. [9] in 1969. He also discovered manganese superoxide dismutase (MnSOD) in 1973 [10]. However, the theory was rendered less compelling 10 years later by the publication of a study demonstrating that superoxide by itself has low levels of activity and reactivity [11]. This was demonstrated by measuring the rate constant of the major radicals in an ascorbic acid solution and ion-balanced solutions, and this counter theory was believed to hold true also in the biologic environment until a few years ago. About 10 years after the presentation of the counter theory, it began to be recognized that nitrogen oxide (NO) plays an important role in the physiologic activities of organisms, and more attention began to be paid to NO [12]. It was subsequently revealed that superoxide readily reacts with NO, which is then converted to highly reactive peroxynitrite [13]. It is well known that in 1998, three distinguished researchers (F. Murat, L. Ignarro, and R. Furchgott) received the Nobel Prize for their achievements in NO research [14].

Emergence of Mitochondria as the Major Mediators of Apoptosis

The development of ROS-related diseases starts with the activation of the protective mechanism against ROS followed by tissue and cell damage [8]. The processes leading to cell death include necrosis and apoptosis [15]. Since the concept of apoptosis was first proposed by Kerr et al. [16] in 1972, the involvement of apoptosis in various diseases has been suggested, and there have been several reports that showed the role of mitochondria in cellular damage and cell death (table 2).

mtDNA is susceptible to oxidative stress and produces mutations at high frequency [17]. Oxidative stress related to the mitochondria has been increasingly observed; for example, aging and mtDNA damage are correlated [18]. Mitochondrial defects have also been observed in neurodegenerative diseases (e.g. Alzheimer's, Parkinson, and ALS) and aging [19].

In the 1970s, it was noted that ROS are generated from mitochondrial electron transport system complexes I and III [20, 21]. Zamzami et al. [22] revealed that mitochondrial membrane potential and apoptosis are correlated. The mitochondrion had been studied and recognized solely as the energy-producing factory of the cell until its direct involvement in apoptosis was demonstrated in 1996 [23]. This great achievement was accomplished by a group of investigators from Emory University in the US. They demonstrated that cytochrome c in the mitochondria serves as an apoptosis-inducing factor. This discovery was made possible by grinding as much as 40 kg of cells, fractionating the homogenate, and searching for the substance that induces apoptosis. Thus, we were confronted with the startling new fact that mitochondria

Table 2. Evidence for the relationship between mitochondria and oxidative stress

a. Mitochondria and Oxidative Stress 1

1. ROS are generated from mitochondrial electron transport system complex I and III [20, 21].
2. Aging and mtDNA damage are correlated [18].
3. Mitochondrial membrane potential and apoptosis are correlated [22].
4. Mitochondrial defects in neurodegenerative diseases (Alzheimer's, Parkinson's, ALS etc.) and aging [19].
5. Cytochrome c, which is located in mitochondria as a part of cellular energy producer in electron transportation system, is released from mitochondria and works as a direct inducer of apoptosis [23, 24].
6. Bcl-2, which protects against apoptosis, obstructs release of cytochrome c from mitochondria [24].
7. mtDNA is susceptible to oxidative stress and produces high frequency of mutations [17].

b. Mitochondria and Oxidative Stress 2

8. Mitochondrial MnSOD protects against oxidative cell death – Mitochondrial ROS regulate apoptosis (1) [25].
9. Mitochondrial phospholipid hydroperoxide glutathione peroxidase protects against oxidative cell death – Mitochondrial ROS regulate apoptosis (2) [27].
10. Mitochondrial apoptosis-inducing factor induces apoptosis [28].
11. Mitochondrial procaspase regulates apoptosis [29, 30].
12. ROS, lipid peroxidation and apoptosis are correlated [61, 62].
13. Impairment of ETC and mtDNA damage induce ROS [35].

control apoptosis. Subsequently, while a number of investigators were attempting to elucidate the function of Bcl-2, a tumor suppressor gene product, the same group of investigators found that Bcl 2 controls the release of cytochrome c out of the mitochondria by adjusting the size of pores on the mitochondrial outer membrane [24]. This paper was identified as the most cited article in *Science* in 1997. In the following year (1998), Majima et al. [25, 26] found that cells transfected with the MnSOD gene are resistant to apoptotic cell death caused by the stimulation of ROS. They demonstrated for the first time that superoxide generated by mitochondria is involved in apoptosis. This finding resulted in the revival of the superoxide theory. In addition, Arai et al. [27] showed that mitochondrial phospholipid hydroperoxide glutathione peroxidase protects against oxidative injuries. The publication of these reports accelerated the elucidation of the relationship between mitochondria and apoptosis. Apoptosis is accompanied by a decrease or change in mitochondrial membrane potential and an increase in intracellular calcium concentration. This form of cell death is now referred to as mitochondria-mediated cell death. Besides cytochrome c, an apoptotic precursor, mitochondria also contain apoptosis-inducing factor [28] and caspase precursors [29, 30], and cell death is induced through these factors. How do mitochondria produce ROS?

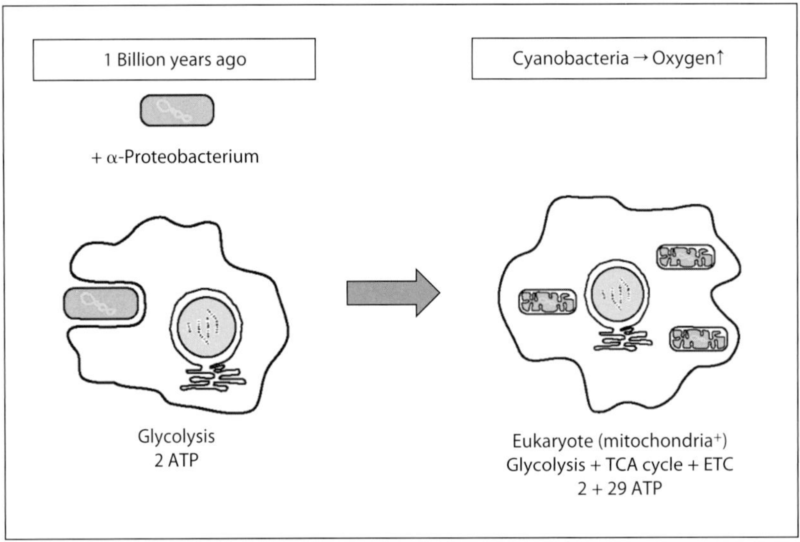

Fig. 1. Schema of 'birth' of mitochondria and consequential oxidative phosphorylation as a result of symbiosis. Oxidative phosphorylation, which is conducted by the ETC, produces 31 ATP instead of only 2 ATP produced by glycolysis. Amended from Gray et al. [31].

Mitochondrial Electron Transport Chain and Superoxide Production

The origin of the mitochondrion is believed to be a bacterium that entered a cell and began a symbiotic relationship about a billion years ago (fig. 1) [31]. The emergence of the mitochondrion resulted in increased production of ATP (from only 2 ATP molecules produced by the glycolytic pathway to 31 molecules) by the ETC in the mitochondrial inner membrane in the presence of oxygen [32]. The ETC consists of complexes I–IV, ATP synthase, and ANP translocator. The mitochondrion also has DNA, which encodes 13 of these proteins. The mtDNA contains codon sequences different from those of the nuclear DNA. In the mitochondrial ETC, while oxidation/reduction of electrons occurs repeatedly, hydrogen ions are transported into the intermembrane space and back to the matrix, ADP is converted into ATP using the produced energy, and the produced ATP is transported into the intermembrane space (fig. 2). This complicated 'bio-machine' serves as the largest energy-producing organelle of the cell, but is an imperfect machine and thus causes 'leakage' of electrons. This leakage most frequently occurs from complexes I and III. The leakage of 2–3% of electrons occurs even in a normal state, and superoxide is considered to be generated from these leaked electrons [19, 21, 33, 34]. More specifically, these leaked electrons are captured by oxygen molecules, and these oxygen molecules become superoxide. With an average adult daily oxygen consumption of about 250 g, the leakage of 2–3% of electrons makes the mitochondrion the major source of superoxide in the cell. The mitochondrion contains a specific enzyme called MnSOD. As can be easily imagined,

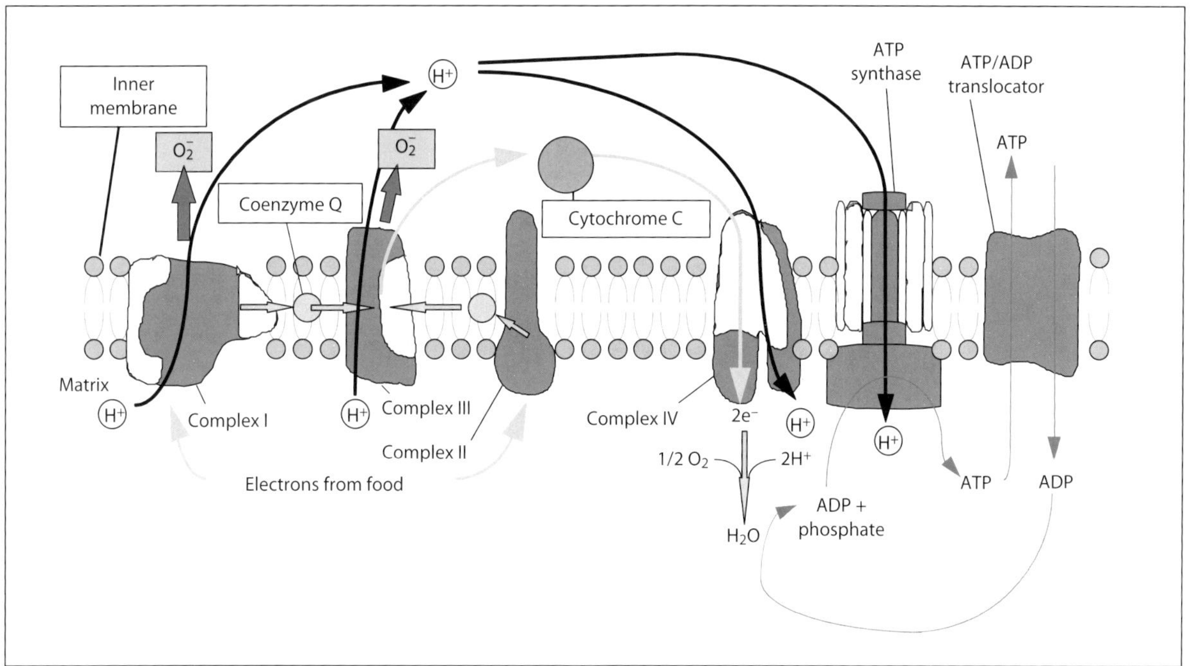

Fig. 2. Schema of the ETC. The ETC is composed of around 100 proteins in total, 4 complexes (complex I–IV), ATP synthase (complex V), and ANT translocator. The yellow areas indicate the proteins that are encoded by the mt genome.

this enzyme plays an important role in capturing superoxide. There had been controversy until a few years ago over the origin of intracellular ROS that play a major role in active oxygen stress. We used imaging techniques to demonstrate that most of the intracellular ROS are generated from mitochondria in intact cells (fig. 3) [35]. It is noted that ROS are located mainly in the mitochondrion, and lipid peroxidation (as determined by HNE detection) is found in a similar location to the ROS.

ROS Generation by Mitochondria in Cells with Impaired ETC

Inhibitors of the ETC block ATP synthesis, although the complex subunit function remains reduced after the addition of ETC substrates such as succinate [35–37]. It has been suggested that oxygen radicals are produced by the ETC when electron flow is interrupted by biochemical inhibitors [20, 38–41]. When the ETC inhibitors rotenone, 3-nitropropionic acid, thenoyltrifluoroacetone, antimycin A, and NaCN (which block complexes I, II, III, and IV) were used, an increase in the generation of mitochondrial ROS was observed in 143B, HeLacot, and mtDNA-lacking 143B-ρ0 cells, as determined by a newly developed fluorescent probe, 2-[6-(4′-hydroxy)phenoxy-

3*H*-xanthen-3-on-9-yl]benzoic acid (HPF) [42]. In the 143B-ρ0 cells, the ETC must function at a lower level than in the parent cells, because of the absence of the 13 mitochondrial-encoded subunits, which are located in complexes I (7 subunits), III (1 subunit), and IV (4 subunits).

Indo et al. [35] demonstrated that more ROS are generated in the electron transport systems of impaired cells. In addition, transient transfection with the MnSOD gene results in reduced ROS generation, indicating that the ROS detected using HPF reflects superoxide generated by mitochondria. Furthermore, most HPF fluorescence was located in mitochondria in every fluorescence image (fig. 3). These results indicate that mitochondria are the major source of intracellular free radicals, and that impairment of the electron transport system could increase mitochondrial generation of ROS.

Reactive Oxygen Species Generation by Mitochondria in Cells with Mitochondrial DNA Damage

mtDNA damage causes a variety of diseases, including neurodegenerative diseases, and contributes to aging [19, 43]. Aging-dependent increases in mtDNA defects [18, 44–46], and the importance of mtDNA defects are also recognized for neurologic disorders, such as Alzheimer's disease, Parkinson's disease, and ALS [4, 7, 47]. The 'common' 4,977-bp deletion of the mtDNA is known to cause Kearns-Sayre syndrome, which is a progressive and ultimately fatal human encephalomyopathy.

More than 90% of the oxygen consumed by mammalian cells is utilized in mitochondria, and as described previously, up to 4% of this oxygen is transformed into ROS [48]. It has been reported that superoxide is generated from complex I [49], complex III [50], and complex II [for review, see 51]. Guidot et al. [52] reported that the ETC is a major source of $O_2^{\bullet-}$ in vivo, and that the principal site of this $O_2^{\bullet-}$ production is proximal to the cytochrome-c-oxidase complex, although Adam-Vizi [53] concluded that sources other than the ETC might also contribute to $O_2^{\bullet-}$ production. Our results demonstrated that mitochondria are clearly a major source of intracellular ROS (fig. 3) [35].

The presence of ETC inhibitors and a lack of mtDNA would create similar environmental conditions and produce similarly poor functioning of the ETC. A significantly greater ROS level was observed in 143B-ρ0 cells than in 143B (parent) and 87wt (cybrid ρ+ cells), both of which contain normal mitochondria. Similarly, HeLacot cells in which there was >50% heteroplasmy with respect to the 4,774-bp common deletion [54] generated more ROS than the parent HeLacot cells. In 143B-ρ0 cells, the 13 subunits of the ETC are lacking, namely, ND1, ND2, ND3, ND4, ND4L, ND5, and ND6 (complex I); cytochrome b (complex III); COX I, COX II, and COX III (complex IV), and ATPase 6 and ATPase 8 (complex V; fig. 2). HeLacot cells with the common 4,977-bp deletion (BH3.12) lack ND3, ND4, ND4L, and a

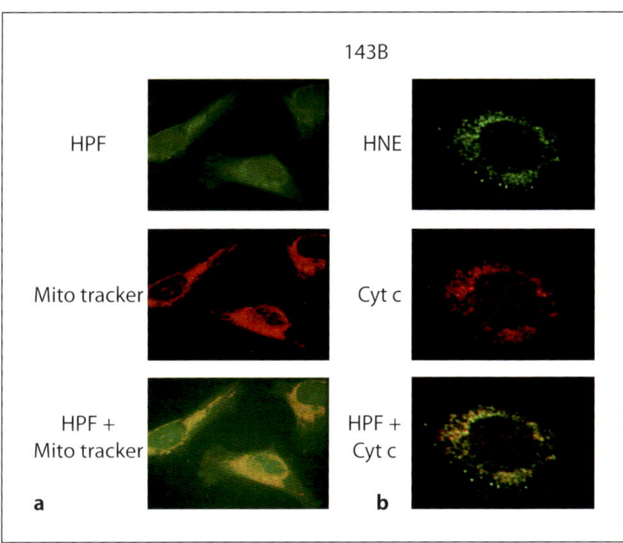

Fig. 3. Most intracellular ROS are generated by mitochondria, and lipid peroxidation occurs around mitochondria. **a** Representative images of live 143B cells in which ROS were visualized using HPF. To examine the location of mitochondria, the same cells were stained with MitoTracker Red CMXRos. Merged double images of HPF and MitoTracker fluorescence were constructed to identify ROS in mitochondria. Most ROS were localized in mitochondria, as shown by the yellow color (green plus red) in the merged images for both cell types. **b** 143B cells were fixed and a major lipid peroxidation product, HNE, was immunocytochemically stained. To localize HNE (green color), double staining was performed with anti-cytochrome c (Cyt c), which localized in the mitochondria (red color). Most HNE was localized in the mitochondria, as observed in the merged image (yellow color). Amended from Indo et al. [35].

part of NAD5 (complex I), and ATPase 6 and a part of ATPase 8 (complex V; fig. 2). These cells must survive using glycolysis as the exclusive mode of ATP production. The question of whether the functions and electron flow of the ETC in 143B-ρ0 and BH3.12 cells are completely eliminated is controversial. If electron flow still existed in the ETC of mtDNA-deleted and mtDNA-depleted cells, ROS generation would occur.

Miranda et al. [55] found higher levels of ROS generation and upregulation of the nuclear-encoded mitochondrial biogenesis genes NRF-1 and Tfam in ρ0 cells than in the parental cells. Marchetti et al. [56] also found that ρ0 cells generated ROS. In ρ0 cells, each remaining complex subunit, which reduced species of ubiquinone or flavosemiquinone intermediates, may increase $O_2^{\bullet -}$ because of auto-oxidation or destabilization due to incomplete assembly of the complex [57]. Furthermore, electron flow may be supported by succinate dehydrogenase (complex II) of which all subunits remain intact in ρ0 cells because they are the products of genes encoded in the nucleus. It has been reported that cells harboring the mitochondrial common deletion also generate more intracellular ROS [58].

Transient MnSOD transfection suppressed intracellular ROS generation in 143B, 143B-ρ0, and 87wt cells. Production of ROS from the ETC results in oxidative stress in cells and causes apoptotic cell death [59, 60]. MnSOD is an essential enzyme, which scavenges superoxide in mitochondria [10]. The biologic importance of MnSOD has been clearly demonstrated in many studies: expression of MnSOD is essential for the survival of aerobic life and the development of cellular resistance to toxicity mediated by oxygen radicals [25, 61, 62]. Transfection of the MnSOD gene into cells resulted in a reduction in ROS generation, subsequent lipid peroxidation, and apoptosis; thus, there is correlation among them. Interference with the ETC or mtDNA damage increases the generation of ROS from mitochondria and causes lipid peroxidation and subsequent apoptosis, indicating that mitochondria are the primary sites of cellular oxidative injuries.

References

1 Halliwell B, Gutterridge JMC: Free radicals, other reactive species and disease; in Halliwell B, Gutterridge JMC (eds): Free Radicals in Biology and Medicine. Oxford, Oxford University Press, 1999, pp 617–784.

2 Halliwell B, Gutterridge JMC: Ageing, nutrition, disease and therapy: a role for antioxidants? in Halliwell B, Gutterridge JMC (eds): Free Radicals in Biology and Medicine. Oxford, Oxford University Press, 1999, pp 784–859.

3 McLellan ME, Kajdasz ST, Hyman BT, Bacskai BJ: In vivo imaging of reactive oxygen species specifically associated with thioflavine S-positive amyloid plaques by multiphoton microscopy. J Neurosci 2003;23:2212–2217.

4 Ikebe S, Tanaka M, Ohno K, Sato W, Hattori K, Kondo T, Mizuno Y, Ozawa T: Increase of deleted mitochondrial DNA in the striatum in Parkinson's disease and senescence. Biochem Biophys Res Commun 1990;170:1044–1048.

5 Beal MF: Mitochondria, oxidative damage, and inflammation in Parkinson's disease. Ann NY Acad Sci 2003;991:120–131.

6 Hoglinger GU, Carrard G, Michel PP, Medja F, Lombes A, Ruberg M, Friguet B, Hirsch EC: Dysfunction of mitochondrial complex I and the proteasome: interactions between two biochemical deficits in a cellular model of Parkinson's disease. J Neurochem 2003;86:1297–1307.

7 Wiedemann FR, Manfredi G, Mawrin C, Beal MF, Schon EA: Mitochondrial DNA and respiratory chain function in spinal cords of ALS patients. J Neurochem 2002;80:616–625.

8 Halliwell B, Gutterridge JMC: Why is superoxide cytotoxic?; in Halliwell B, Gutterridge JMC (eds): Free Radicals in Biology and Medicine. Oxford, Oxford University Press, 1999, pp 129–134.

9 McCord JM, Fridovich I: Superoxide dismutase. An enzymic function for erythrocuprein (hemocuprein). J Biol Chem 1969;244:6049–6055.

10 Weisiger RA, Fridovich I: Mitochondrial superoxide dismutase. J Biol Chem 1973;248:4793–4796.

11 Sawyer DT, Valentine JS: How super is superoxide? Acc Chem Res 1981;14:393–340.

12 Palmer RM, Ferrige AG, Moncada S: Nitric oxide release accounts for the biological activity of endothelium-derived relaxing factor. Nature 1987;327:524–526.

13 Beckman JS, Beckman TW, Chen J, Marshall PA, Freeman BA: Apparent hydroxyl radical production by peroxynitrite: implications for endothelial injury from nitric oxide and superoxide. Proc Natl Acad Sci U S A 1990;87:1620–1624.

14 http://nobelprize.org/nobel_prizes/medicine/laureates/1998/.

15 Halestrap A: Biochemistry: a pore way to die. Nature 2005;434:578–579.

16 Kerr JF, Wyllie AH, Currie AR: Apoptosis: a basic biological phenomenon with wide-ranging implications in tissue kinetics. Br J Cancer 1972;26:239–257.

17 Wei YH: Oxidative stress and mitochondrial DNA mutations in human aging. Proc Soc Exp Biol Med 1998;217:53–63.

18 Linnane AW, Marzuki S, Ozawa T, Tanaka M: Mitochondrial DNA mutations as an important contributor to ageing and degenerative diseases. Lancet 1989;333:642–645.

19 Wallace DC: Mitochondrial DNA in aging and disease. Sci Am 1997;277:40–47.
20 Boveris A, Cadenas E: Mitochondrial production of superoxide anions and its relationship to the antimycin insensitive respiration. FEBS Lett 1975;1:311–314.
21 Takeshige K, Minakami S: NADH- and NADPH-dependent formation of superoxide anions by bovine heart submitochondrial particles and NADH-ubiquinone reductase preparation. Biochem J 1979;180:129–135.
22 Zamzami N, Marchetti P, Castedo M, Decaudin D, Macho A, Hirsch T, Susin SA, Petit PX, Mignotte B, Kroemer G: Sequential reduction of mitochondrial transmembrane potential and generation of reactive oxygen species in early programmed cell death. J Exp Med 1995;182:367–377.
23 Liu X, Kim CN, Yang J, Jemmerson R, Wang X: Induction of apoptotic program in cell-free extracts: requirement for dATP and cytochrome c. Cell 1996;86:147–157.
24 Yang J, Liu X, Bhalla K, Kim CN, Ibrado AM, Cai J, Peng TI, Jones DP, Wang X: Prevention of apoptosis by Bcl-2: release of cytochrome c from mitochondria blocked. Science 1997;275:1129–1132.
25 Majima H, Oberley TD, Furukawa K, Mattson MP, Yen H-C, Szweda LI, St Clair DK: Prevention of mitochondrial injury by manganese superoxide dismutase reveals a primary mechanism for alkaline-induced cell death. J Biol Chem 1998;273:8217–8224.
26 Majima HJ, Indo HP, Tomita K, Suenaga S, Motoori S, Kato H, Yen H-C, Ozawa T: Intracellular oxidative stress caused by ionizing radiation; in Singh KK (ed): Oxidative Stress, Disease and Cancer. London, Imperial College Press, 2006, pp 61–83.
27 Arai M, Imai H, Koumura T, Yoshida M, Emoto K, Umeda M, Chiba N, Nakagawa Y: Mitochondrial phospholipid hydroperoxide glutathione peroxidase plays a major role in preventing oxidative injury to cells. J Biol Chem 1999;274:4924–4933.
28 Susin SA, Lorenzo HK, Zamzami N, Marzo I, Snow BE, Brothers GM, Mangion J, Jacotot E, Costantini P, Loeffler M, Larochette N, Goodlett DR, Aebersold R, Siderovski DP, Penninger JM, Kroemer G: Molecular characterization of mitochondrial apoptosis-inducing factor. Nature 1999;397:441–446.
29 Susin SA, Lorenzo HK, Zamzami N, Marzo I, Brenner C, Larochette N, Prévost MC, Alzari PM, Kroemer G: Mitochondrial release of caspase-2 and -9 during the apoptotic process. J Exp Med 1999;189:381–394.
30 Suzuki A, Tsutomi Y, Yamamoto N, Shibutani T, Akahane K: Mitochondrial regulation of cell death: mitochondria are essential for procaspase 3-p21 complex formation to resist Fas-mediated cell death. Mol Cell Biol 1999;19:3842–3847.
31 Gray MW, Burger G, Lang BF: Mitochondrial evolution. Science 1999;283:1476–1481.
32 Salway JG: Oxidation of glucose yields 31 ATP molecules assuming the 'modern' P/O ratios of 2.5 for NADH and 1.5 for FADH2, in Salway JG (ed): Metabolism at a Glance. Malden, Blackwell Publishing, 2004, pp 21.
33 Beyer RE: An analysis of the role of coenzyme Q in free radical generation and as an antioxidant. Biochem Cell Biol 1992;70:390–403.
34 Boveris A, Chance B: The mitochondrial generation of hydrogen peroxide. General properties and effect of hyperbaric oxygen. Biochem J 1973;134:707–716.
35 Indo HP, Davidson M, Yen H-C, Suenaga S, Tomita K, Nishii T, Higuchi M, Koga Y, Ozawa T, Majima HJ: Evidence of ROS generation by mitochondria in cells with impaired electron transport chain and mitochondrial DNA damage. Mitochondrion 2007;7:106–118.
36 Sato N, Wilson DF, Chance B: Two b cytochromes of pigeon heart mitochondria. FEBS Lett 1971;15:209–212.
37 Dutton PL, Erecinska M, Sato N, Mukai Y, Pring M, Wilson DF: Reactions of b-cytochromes with ATP and antimycin A in pigeon heart mitochondria. Biochim Biophys Acta 1972;267:15–24.
38 Cross AR, Jones OT: Enzymic mechanisms of superoxide production. Biochim Biophys Acta 1991;1057:281–298.
39 Zhang JG, Nicholls Grzemski FA, Tirmenstein MA, Fariss MW: Vitamin E succinate protects hepatocytes against the toxic effect of reactive oxygen species generated at mitochondrial complexes I and III by alkylating agents. Chem Biol Interact 2001;138:267–284.
40 Yamagishi S, Edelstein D, Du X, Kaneda Y, Guzman M, Brownlee M: Leptin induces mitochondrial superoxide production and monocyte chemoattractant protein-1 expression in aortic endothelial cells by increasing fatty acid oxidation via protein kinase A. J Biol Chem 2001;276:25096–25100.
41 Muller FL, Liu Y, Van Remmen H: Complex III releases superoxide to both sides of the inner mitochondrial membrane. J Biol Chem 2004;279:49064–49073.
42 Setsukinai K, Urano Y, Kakinuma K, Majima HJ, Nagano T: Development of novel fluorescence probes that can reliably detect reactive oxygen species and distinguish specific species. J Biol Chem 2003;278:3170–3175.

43 Barja G: Free radicals and aging. Trends Neurosci 2004;27:595–600.
44 Corral-Debrinski M, Horton T, Lott MT, Shoffner JM, Beal MF, Wallace DC: Mitochondrial DNA deletions in human brain; regional variability and increase with advanced age. Nat Genet 1992;2:324–329.
45 Hayakawa M, Hattori K, Sugiyama S, Ozawa T: Age-associated oxygen damage and mutations in mitochondrial DNA in human hearts. Biochem Biophys Res Commun 1992;189:979–985.
46 Chomyn A, Attardi G: MtDNA mutations in aging and apoptosis. Biochem Biophys Res Commun 2003;304:519–529.
47 Sheehan JP, Swerdlow RH, Miller SW, Davis RE, Parks JK, Parker WD, Tuttle JB: Calcium homeostasis and reactive oxygen species production in cells transformed by mitochondria from individuals with sporadic Alzheimer's disease. J Neurosci 1997;17: 4612–4622.
48 Chance B, Sies H, Boveris A: Hydroperoxide metabolism in mammalian organs. Physiol Rev 1979;59: 527–605.
49 Kushnareva Y, Murphy AN, Andreyev A: Complex I-mediated reactive oxygen species generation: modulation by cytochrome c and NAD(P)+ oxidation-reduction state. Biochem J 2002;368:545–553.
50 Chen Q, Vazquez EJ, Moghaddas S, Hoppel CL, Lesnefsky EJ: Production of reactive oxygen species by mitochondria: central role of complex III. J Biol Chem 2003;278:36027–36031.
51 Jezek P, Hlavata L: Mitochondria in homeostasis of reactive oxygen species in cell, tissues, and organism. Int J Biochem Cell Biol 2005;37:2478–2503.
52 Guidot DM, McCord JM, Wright RM, Repine JE: Absence of electron transport (Rho0 state) restores growth of a manganese-superoxide dismutase-deficient *Saccharomyces cerevisiae* in hyperoxia. Evidence for electron transport as a major source of superoxide generation in vivo. J Biol Chem 1993;268: 26699–26703.
53 Adam-Vizi V: Production of reactive oxygen species in brain mitochondria: contribution by electron transport chain and non-electron transport chain sources. Antioxid Redox Signal 2005;7:1140–1149.
54 Sancho S, Moraes CT, Tanji K, Miranda AF: Structural and functional mitochondrial abnormalities associated with high levels of partially deleted mitochondrial DNAs in somatic cell hybrids. Somat Cell Mol Genet 1992;18:431–442.
55 Miranda S, Foncea R, Guerrero J, Leighton F: Oxidative stress and upregulation of mitochondrial biogenesis genes in mitochondrial DNA-depleted HeLa cells. Biochem Biophys Res Commun 1999; 258:44–49.
56 Marchetti P, Susin SA, Decaudin D, Gamen S, Castedo M, Hirsch T, Zamzami N, Naval J, Senik A, Kroemer G: Apoptosis-associated derangement of mitochondrial function in cells lacking mitochondrial DNA. Cancer Res 1996;56:2033–2038.
57 Bandy B, Davison AJ: Mitochondrial mutations may increase oxidative stress: implications for carcinogenesis and aging? Free Radic Biol Med 1990;8:523–539.
58 Peng T-I, Yu P-R, Chen J-Y, Wang H-L, Wu H-Y, Wei Y-H, Jou M-J: Visualizing common deletion of mitochondrial DNA-augmented mitochondrial reactive oxygen species generation and apoptosis upon oxidative stress. Biochim Biophys Acta 2006; 1762:241–255.
59 Mattson MP: Apoptosis in neurodegenerative disorders. Nat Rev Mol Cell Biol 2000;1:120–129.
60 Mattson MP, Duan W, Pedersen WA, Culmsee C: Neurodegenerative disorders and ischemic brain diseases. Apoptosis 2001;6:69–81.
61 Motoori S, Majima HJ, Ebara M, Kato H, Hirai F, Kakinuma S, Yamaguchi C, Ozawa T, Nagano T, Tsujii H, Saisho H: Overexpression of mitochondrial manganese superoxide dismutase protects against radiation-induced cell death in the human hepatocellular carcinoma cell line, HLE. Cancer Res 2001;61:5382–5388.
62 Hirai F, Motoori S, Kakinuma S, Tomita K, Indo HP, Kato H, Yamaguchi T, Yen H-C, St Clair DK, Nagano T, Ozawa T, Saisho H, Majima HJ: Mitochondrial signal lacking manganese superoxide dismutase failed to prevent cell death by reoxygenation following hypoxia in a human pancreatic cancer cell line, KP4. Antioxid Redox Signal 2004;6:523–535.

Hideyuki J. Majima
Kagoshima University Graduate School of Medical and Dental Sciences
Department of Oncology and Space Environmental Medicine
Kagoshima 890-8544 (Japan)
Tel. +81 99 275 6270, Fax +81 99 275 6278, E-Mail hmajima@dent.kagoshima-u.ac.jp

Naito Y, Suematsu M, Yoshikawa T (eds): Free Radical Biology in Digestive Diseases.
Front Gastrointest Res. Basel, Karger, 2011, vol 29, pp 23–34

The Nox Family of NADPH Oxidases that Deliberately Produce Reactive Oxygen Species

Hideki Sumimoto[a] · Reiko Minakami[a,b] · Kei Miyano[a]

Departments of [a]Biochemistry and [b]Health Sciences, Kyushu University Graduate School of Medical Sciences, Fukuoka, Japan

Abstract

Reactive oxygen species (ROS) are classically considered as harmful by-products in aerobic metabolism; on the other hand, they are also deliberately produced by the Nox family NADPH oxidases on the membrane of a variety of cells. The human genome encodes seven members of the family: five Nox proteins (Nox1–Nox5) and the two dual oxidases Duox1 and Duox2 that produce superoxide or hydrogen peroxide as the 'true' primary product. Nox/Duox oxidases, except the constitutively active enzyme Nox4, are regulated via intracellular signal transduction. Activation of Nox1–3 requires binding of their stable membrane partner p22*phox* to the soluble protein p47*phox* (or its homologue Noxo1) complexed with the oxidase activator p67*phox* (or its related protein Noxa1), an event that is triggered by stimulus-induced phosphorylation of p47*phox*; the small GTPase Rac is upon cell stimulation converted to the GTP-bound form and subsequently interacts with p67*phox* or Noxa1, leading to oxidase activation. Nox5 as well as Duox1/2 harbors an N-terminal cytoplasmic extension with Ca^{2+}-binding EF-hand motifs and is thus effectively activated by elevation in the intracellular concentration of Ca^{2+}. The phagocyte oxidase Nox2 plays a crucial role in host defense by producing microbicidal ROS during phagocytosis. This oxidase also appears to function in immunoregulation, since genetic ablation of the Nox2 system causes not only severe pyrogenic infections but also non-infectious granulomatous inflammation similar to inflammatory bowel diseases. Both Nox1 and Duox2 are highly expressed in epithelial cells of the gastrointestinal tract, and could be involved in digestive diseases.

As conventionally regarded, a substantial amount of reactive oxygen species (ROS) originates as by-products in aerobic metabolism such as the mitochondrial respiratory chain. On the other hand, there exist enzymes dedicated to ROS generation, examples of which are membrane-integrated Nox family oxidases that produce superoxide anion (O_2^-) or hydrogen peroxide (H_2O_2) as 'true' product in conjunction with NADPH oxidation [1–6]. The human genome contains seven distinct genes encoding

Table 1. Members of the Nox family in humans

	Genetic locus	Amino acids	Abundant expression
Nox1	Xq22	564	colon, VSMC
Nox2	Xp21.1	570	phagocyte, B lymphocyte
Nox3	6q25.1–26	568	inner ear, fetal kidney
Nox4	11q14.2–21	578	kidney, VEC
Nox5	15q22.31	737	testis, spleen, lymph node
Duox1	15q21	1,551	thyroid gland, salivary gland, lung
Duox2	15q21	1,548	thyroid gland, colon

VSMC = Vascular smooth muscle cell; VEC = vascular endothelial cell.

NADPH oxidases of the Nox family (table 1): the five related Nox enzymes (Nox1 through Nox5) that reduce molecular oxygen to superoxide as primary product, and the two distantly-related dual oxidases Duox1 and Duox2 that release hydrogen peroxide instead of superoxide [1–6]. The founding member is Nox2, also known as gp91^{phox}, which is abundantly expressed in professional phagocytes such as neutrophils. The nonphagocytic oxidases Nox1 and Nox3 to be next identified are structurally very close to gp91^{phox}/Nox2, whereas Nox4 and Nox5 are more distantly related in this order [1–6]. The phagocyte oxidase (*phox*) gp91^{phox}/Nox2 plays a crucial role in host defense by producing superoxide as a precursor of microbicidal ROS during phagocytosis. In addition to the role in innate immunity, Nox and Duox oxidases also serve a variety of biological functions, such as signal transduction and biosynthetic processes [1–8]. In this brief review, we summarize molecular structure, function, and regulatory mechanism of the Nox family NADPH oxidases, and describe their (possible) involvement in digestive diseases.

Structure of Nox Family NADPH Oxidases

The phagocyte oxidase gp91^{phox}/Nox2 and its closely related members of the Nox family (Nox1, Nox3, and Nox4) form a heterodimer with the membrane-integrated protein p22^{phox} (fig. 1); both subunits of the dimer are stabilized by complex formation [1–6]. These oxidases comprise two regions of a similar size: the N-terminal half comprising six predicted α-helical transmembrane segments and the C-terminal half of a cytoplasmic domain homologous to ferredoxin–NADP$^+$ reductase (FNR) (fig. 1). On the basis of biochemical and database analyses, it is thought that the third and

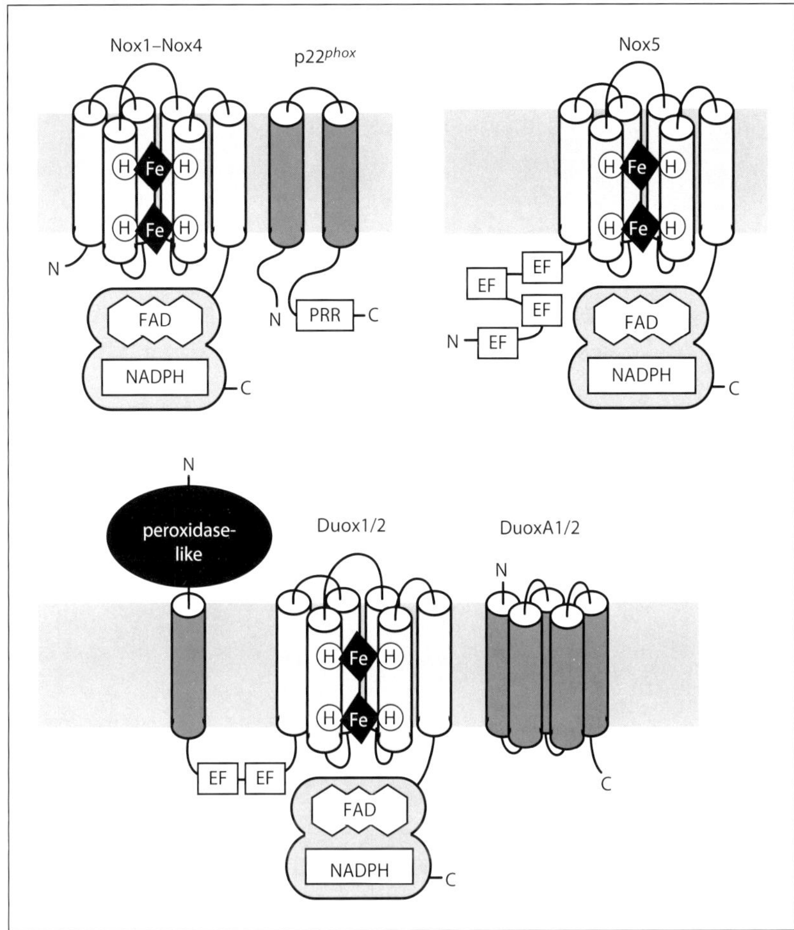

Fig. 1. Structure of human Nox family oxidases. Cylinders represent predicted transmembrane α helices. Nox, NADPH oxidase; PRR, proline-rich region; EF, Ca^{2+}-binding EF-hand motif; Duox1/2, dual oxidase 1 or dual oxidase 2; DuoxA1/2, Duox activator 1 or Duox activator 2.

fifth helices in the N-terminal moiety each contain two invariant histidine residues that coordinate the irons of two nonidentical hemes in the membrane, and that the C-terminal cytoplasmic FNR-homologous domain bears both NADPH- and flavin adenine dinucleotide (FAD)-binding sites [5]. Thus the bipartite structure confers the Nox superdomain that completely transports electrons from NADPH via FAD and two hemes to molecular oxygen, thereby catalyzing superoxide production.

In addition to the Nox superdomain comprising the bis-heme-containing transmembrane region and the FNR homologous moiety, Nox5 is N-terminally extended by a cytoplasmic region that harbors four Ca^{2+}-binding EF-hand motifs [1–6] (fig. 1). In contrast to other Nox and Duox oxidases, Nox5 appears to be stable on its own and effectively targeted to plasma membranes without any associated proteins:

a membrane-bound partner for Nox5, like p22^{phox} for other Nox proteins or DuoxA (Duox activator) for Duox enzymes, has not been thus far identified.

The N-terminal extension of Duox subfamily oxidases contains an extracellular peroxidase-like domain followed by a membrane-spanning segment and two EF-hand motifs [1–6] (fig. 1). A recent study has shown that the peroxidase-like ectodomain of human Duox1 exhibits no intrinsic peroxidase activity [9], as expected by the absence of key active site residues in most of Duox enzymes [2, 5, 6]; this indicates that Duox is not a real 'dual' oxidase. Duox1 and Duox2 are thought to be complexed on plasma membranes with their respective partners DuoxA1 and DuoxA2, both of which harbor five transmembrane α-helices (fig. 1) [10, 11]. Without these partners, Duox oxidases fail to undergo maturation and thus are retained in the endoplasmic reticulum. Once complexed with DuoxA, Duox is transported as a heterodimer to cell surfaces [10, 11]. At plasma membranes, Duox releases hydrogen peroxide without forming detectable amounts of superoxide, although Duox has a C-terminal region homologous to the superoxide-producing Nox superdomain. The reason for the discrepancy is presently unknown: it seems possible that the N-terminal extracellular region may confer the unique property of hydrogen peroxide production, but a recent study has shown that the ectodomain lacks significant superoxide dismutase activity that converts superoxide to hydrogen peroxide [9].

Expression, Function, and Regulation of gp91^{phox}/Nox2

gp91^{phox}/Nox2 is expressed abundantly in professional phagocytes such as neutrophils and monocytes/macrophages, and, to a lesser extent, in a variety of cell type including B lymphocytes and vascular wall cells (endothelial cells, smooth muscle cells, and adventitial fibroblasts). The importance of the phagocyte oxidase in host defense is exemplified by recurrent and life-threatening infections that occur in patients with chronic granulomatous disease (CGD), whose phagocytes genetically lack the superoxide-producing activity [1–6].

Maturation and stabilization of gp91^{phox}/Nox2 requires complex formation at membranes with p22^{phox} [1–6], which also serves as an anchor for the oxidase-activating protein p47^{phox} as described below. The phagocyte oxidase gp91^{phox}/Nox2 is dormant in resting cells, but becomes activated to produce superoxide during phagocytosis or in response to soluble agonists such as chemoattractants.

Activation of gp91^{phox}/Nox2 requires soluble oxidase proteins (the SH3 domain-containing proteins p47^{phox}, p67^{phox}, and p40^{phox}, and the small GTPase Rac; see table 2) as well as stimulus-induced signal transduction to these proteins (phosphorylation of p47^{phox} and conversion of Rac to the GTP-bound form). The significance of these proteins becomes evident by the fact that CGD is caused not only by a defect in the gp91^{phox}/Nox2 gene but also by a defect in any of the genes for p22^{phox} and the soluble activating proteins [1–6, 12]. In the cytosol of resting cells, p47^{phox} forms a

Table 2. Regulation of the Nox family oxidases

	Membrane partner	Soluble activating protein	Involvement of Rac	Direct activation by Ca^{2+}
Nox1	p22*phox*	Noxo1, Noxa1	+	–
Nox2	p22*phox*	p67*phox*, p47*phox*, p40*phox*[1]	+	–
Nox3	p22*phox*	p67*phox*, p47*phox*, Noxo1	+	–
Nox4	p22*phox*	–	–	–
Nox5	–	–	–	+
Duox1	DuoxA1	–	–	+
Duox2	DuoxA2	–	–	+

[1] p40*phox* is dispensable for activation of Nox2, but facilitates membrane translocation of p67*phox* and p47*phox*, thereby enhancing superoxide production by Nox2.

ternary complex with p67*phox* and p40*phox*, and the small GTPase Rac exists as a dimer with RhoGDI (fig. 2). The bis-SH3 and PX domains of p47*phox* are kept in a state incapable of binding to their respective membrane targets p22*phox* and phosphatidylinositol 3,4-bisphosphate (PI(3,4)P_2) via autoinhibitory region (AIR)-mediated intramolecular interactions (fig. 2). Upon cell stimulation, p47*phox* undergoes phosphorylation on AIR, which enables the bis-SH3 and PX domains to interact with the proline-rich region (PRR) of p22*phox* and PI(3,4)P_2, respectively; the phosphorylation may be catalyzed by multiple kinases such as protein kinase C (PKC). The two induced interactions are both crucial for membrane translocation of the p47*phox*–p67*phox*–p40*phox* complex. The Ile-152-containing region of p47*phox*, N-terminal to the bis-SH3 domain, subsequently plays an essential role in gp91*phox*/Nox2 activation (fig. 2) [13]. On the other hand, Rac dissociates from RhoGDI and localizes to membranes in the GTP-bound state, thereby interacting with p67*phox* in the ternary complex translocated independently (fig. 2).

In addition to the N-terminal Rac-binding domain comprising four tetratricopeptide repeat (TPR) motifs, p67*phox* is composed of the N-terminal SH3 domain, the PB1 domain, and the C-terminal SH3 domain that binds to the PRR of p47*phox* (fig. 3a). In contrast to p47*phox*, p67*phox* appears to adopt an elongated conformation with no or little apparent associations between the domains [14]. Although p40*phox* is dispensable for gp91*phox*/Nox2 activation, it constitutively associates with p67*phox* via the PB1–PB1 interaction [15] to facilitate translocation to membranes, especially phagosomal membranes [1–6]. At membranes GTP-bound Rac interacts with p67*phox*, which may enable the activation domain, C-terminal to the Rac-binding domain, to directly interact with gp91*phox*/Nox2 [1–6], culminating in superoxide

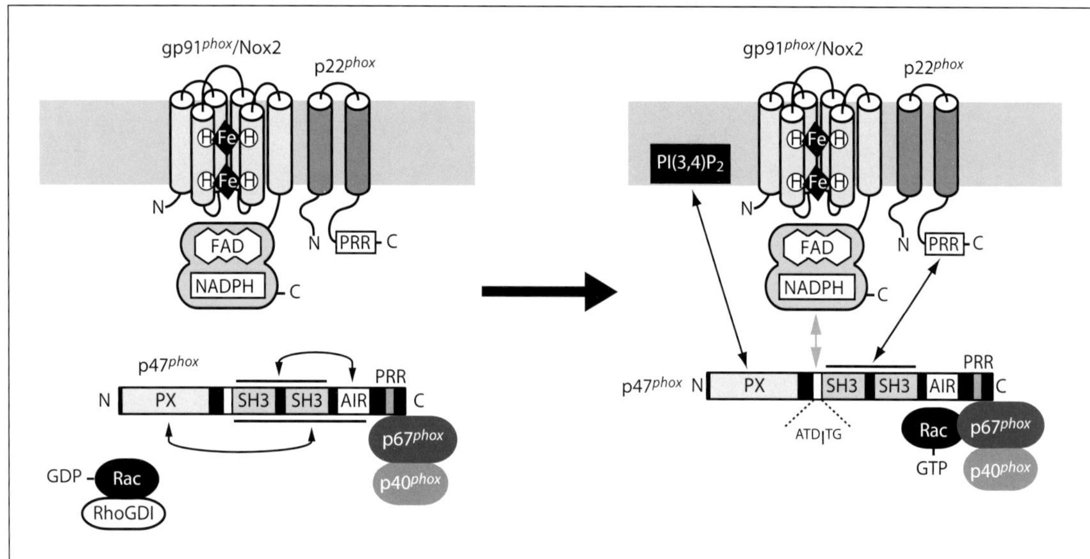

Fig. 2. Stimulus-induced conformational change of p47*phox* for gp91*phox*/Nox2 activation. In the cytosol of resting cells, p47*phox* forms a ternary complex with p67*phox* and p40*phox*, whereas the small GTPase Rac is complexed with RhoGDI. The bis-SH3 and PX domains of p47*phox* are kept in a state incapable of binding to their target molecules via autoinhibitory region (AIR)-mediated intramolecular interactions. Upon cell stimulation, the ternary complex as well as Rac translocates to the membrane where gp91*phox*/Nox2 exists. p47*phox* undergoes a stimulus-induced conformational change, which enables the bis-SH3 and PX domains to interact with the proline-rich region (PRR) of p22*phox* and phosphatidylinositol 3,4-bisphosphate (PI(3,4)P_2), respectively. After translocation, the Ile-152-containing region N-terminal to bis-SH3 domain plays an essential role in gp91*phox*/Nox2 activation. On the other hand, Rac dissociates from RhoGDI and localizes to membranes as the GTP-bound form.

production (fig. 3a). Among small GTPases, Rho family proteins such as Rac solely bear an extra helix called 'insert region'; this region of Rac, however, is not involved in gp91*phox*/Nox2 activation [16]. In addition, the N-terminal SH3 domain of p67*phox* seems to enhance the interaction of this activator protein with gp91*phox*/Nox2 [14, 17] (fig. 3a).

Expression and Regulation of Nox1

The first mammalian nonphagocytic oxidase to be cloned is Nox1, which exists most abundantly in epithelial cells of the colon, but also at lower levels in various types of cell including vascular smooth muscle cells [18]. Although the role of Nox1 in the colon still remains unclear, it may participate in host defense. On the other hand, Nox1 in the vasculature is known to play a role in angiotensin II-based regulation of vascular tone [4–6, 18].

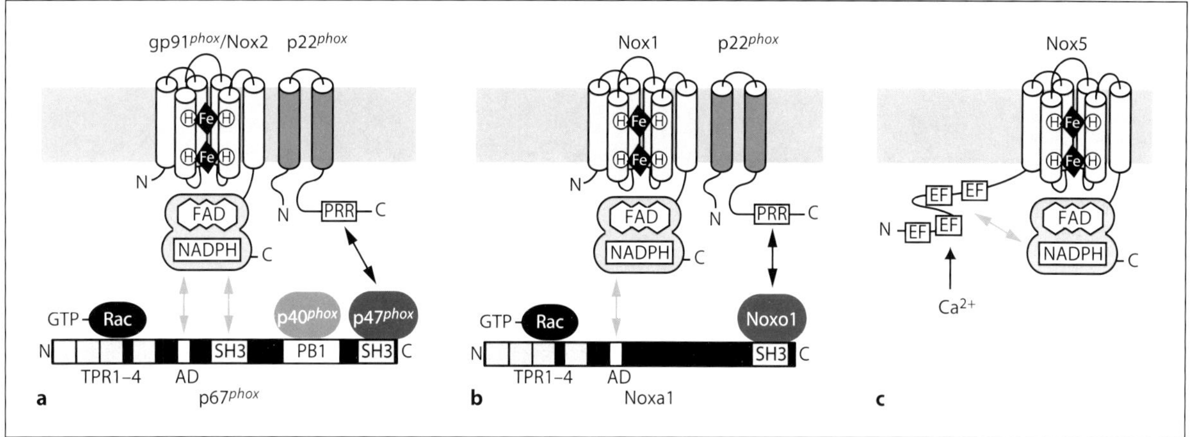

Fig. 3. Activation of gp91^{phox}/Nox2, Nox1, and Nox5. (**a**) Role of p67^{phox} in gp91^{phox}/Nox2 activation. p67^{phox} binds to Rac, p40^{phox}, and p47^{phox} via the TPR, PB1, and C-terminal SH3 domains, respectively. These bindings likely allow p67^{phox} to interact with gp91^{phox}/Nox2 via the AD and the N-terminal SH3 domain, leading to superoxide production. (**b**) Role of Noxa1 in Nox1 activation. Noxa1 binds to Rac and Noxo1 via the TPR and SH3 domains, respectively. The AD of Noxa1 probably interacts with Nox1, leading to superoxide production. (**c**) Role of Ca^{2+} in Nox5 activation. Ca^{2+} binds to the EF-hand motifs in the N-terminal cytoplasmic region of Nox5. The Ca^{2+}-bound EF-hand motifs are then proposed to interact intramolecularly with the C-terminal FNR-homologous domain of Nox5, leading to superoxide production. TPR1–4, tetratricopeptide repeats 1 through 4; AD, activation domain.

Superoxide production by Nox1 as well as that by gp91^{phox}/Nox2 requires the membrane-integral partner p22^{phox} [1–6]. Nox1 activation depends on interactions with the cytosolic regulatory proteins designated Nox organizer 1 (Noxo1) and Nox activator 1 (Noxa1), which are homologous to p47^{phox} and p67^{phox}, respectively [1–6] (fig. 3b). As with gp91^{phox}/Nox2, full Nox1 activity is dependent on Rac [5, 6] (table 2). Noxa1 as well as p67^{phox} harbors the Rac-binding TPR domain, the activation domain, and the C-terminal SH3 domain for interacting with Noxo1 (fig. 3b), which is consistent with the significance of Rac and Noxo1 in Nox1 activation. The reason why the N-terminal SH3 and PB1 domains are absent from the nonphagocytic activator Noxa1 may be explicable by the findings that the N-terminal SH3 domain of p67^{phox} enhances superoxide production by gp91^{phox}/Nox2 but not by Nox1 [17] and that Noxa1 does not require the PB1 partner p40^{phox}, a protein crucial solely for translocation to phagosomal membranes [5, 6]. In the presence of Noxo1 and Noxa1, Nox1 is capable of producing superoxide in an agonist-independent manner, which is consistent with the finding that, unlike p47^{phox}, Noxo1 constitutively associates with p22^{phox} because of the lack of AIR in this organizer protein [1–6]. Superoxide production by Nox1 can be further enhanced by cell stimulation with agonists; the enhancement is likely mediated at least in part via conversion of Rac to the GTP-bound form, since GTP-bound Rac solely binds to Noxa1 and participates in Nox1 activation.

Expression, Function, and Regulation of Nox3

Nox3, expressed in the inner ear and the fetal kidney, is closely related to Nox1 and $gp91^{phox}$/Nox2, and forms a stable heterodimer with $p22^{phox}$ as well [2–6]. It is thus reasonable that Nox3 activity can be strongly facilitated by the $p22^{phox}$-interacting proteins $p47^{phox}$ and Noxo1, whereas it is also upregulated by Rac in the presence of $p67^{phox}$ or Noxa1 [2–6] (table 2). In contrast to Nox1 and $gp91^{phox}$/Nox2, however, Nox3 has an ability to produce a small but considerable amount of superoxide in the absence of an organizer or an activator [2–6]. In mice, Nox3 plays a crucial role in the genesis of otoconia, tiny mineralized structures that are required for sensing balance and gravity [2–6]. Consistent with this, mice with deletion of $p22^{phox}$, an integral membrane protein that stabilizes not only Nox2 but also Nox3 by forming a heterodimer, have been reported to exhibit a compound phenotype consisting of a CGD-like immune defect and a balance disorder resulting from the aberrant development of gravity-sensing organs [19]. On the other hand, no report has described involvement of hereditary disorder in perception of balance and gravity in patients with $p22^{phox}$-deficient CGD. The significance of human Nox3 [20] should be clarified in the future studies.

Expression and Regulation of Nox4

Nox4 is present in a variety of cells, especially in epithelial cells of the adult and fetal kidney, vascular endothelial cells, and, to a lesser extent, cardiac myocytes; this oxidase is known to be upregulated by transforming growth factor-β in all cell types tested [18]. Like Nox1 to Nox3, Nox4 also forms a complex with $p22^{phox}$. However the feature of $p22^{phox}$ does not appear to be conserved in the Nox4-based oxidase [21], which may explain the reason why Nox4 is not regulated by $p22^{phox}$-binding oxidase organizers, oxidase activators, or Rac. All the data presently available indicate that Nox4 exists in a constitutively active state [22–24]. It is thus considered that the expression level of Nox4 essentially determines the amount of ROS produced in cells. Uncontrolled expression of Nox4 appears to result in excessive production of ROS, which exerts deleterious effects on the cells: for instance, it has recently been shown that upregulation of Nox4 in the heart by hypertrophic stimuli and aging induces oxidative stress and cardiac dysfunction [25]. The physiological role of Nox4, however, still remains elusive. Recent studies have suggested that Nox4 may produce hydrogen peroxide more directly than superoxide [21–23]. This could be especially important when a role of Nox4 in the vasculature is considered, since superoxide but not hydrogen peroxide is capable of inactivating the vasodilator nitric oxide.

Expression and Regulation of Nox5

The Nox5 gene is present in a wide variety of mammals, but has been lost in the lineage of rodents [5], which limits experimental models to cultured cells and isolated tissues. In humans, Nox5 is highly expressed in T and B lymphocytes of spleen and lymph nodes, and also in the sperm precursors of testis, although the role for Nox5 is presently unknown. Nox5 is inactive in resting cells, but immediately produces superoxide upon cell stimulation.

In contrast to Nox1 through Nox4, Nox5 does not associate with p22^{phox}. It is thus reasonable that Nox5 activation is independent of p22^{phox}-binding oxidase organizers, oxidase activators, or Rac. Instead Nox5 is activated when the cytoplasmic concentration of Ca^{2+} becomes elevated, as shown by the observation that the Ca^{2+}-ionophore ionomycin effectively elicits superoxide production in Nox5-expressing cells [1–6]. This agrees with the presence of four Ca^{2+}-binding EF-hand motifs in the N-terminal cytoplasmic region of Nox5, which is not found in Nox1 through Nox4 (fig. 1). The Nox5 N-terminal region in the Ca^{2+}-bound state is proposed to interact with the C-terminal FNR-related domain (fig. 3c); the interaction may induce a conformational change that allows electron transfer from NADPH, leading to superoxide production [26]. Sensitivity to Ca^{2+} is likely increased by the Ca^{2+}-binding protein calmodulin and by phosphorylation of Nox5 by PKC: cell stimulation with phorbol myristate acetate, a potent activator of PKC, strongly facilitates superoxide production catalyzed by Nox5 [5, 6].

Expression, Function, and Regulation of Duox1 and Duox2

Although both Duox1 and Duox2 were first identified in the thyroid gland, Duox2 is more highly expressed in the thyroid than Duox1 and thus thought to be mainly responsible for thyroid hormone synthesis. This is consistent with the finding that several mutations in Duox2 cause congenital hypothyroidism; on the other hand, no such mutations have been found in Duox1. The localization of Duox oxidases is not exclusive to the thyroid gland, but they are also expressed at the apical side of surface epithelia exposed to microbes; Duox1 of the airways and Duox2 of the salivary gland and digestive tract. Duox oxidases release hydrogen peroxide on apical surfaces of epithelial cells in mucosal tissues to support dedicated extracellular hemoperoxidases such as thyroperoxidase (TPO) and lactoperoxidase (LPO). In thyrocytes, hydrogen peroxide produced by Duox2 activates TPO, which catalyzes thyroid hormone synthesis. In the lung and salivary glands, Duox provides hydrogen peroxide to LPO, which converts thiocyanate anions into the microbial oxidant hypothiocyanate and thus may serve as an integral part of the innate host defense system and inflammatory responses at mucosal surfaces [6, 7].

Basal activity of both Duox isozymes depends on Ca^{2+} and paired EF-hand motifs, and elevation in the cytoplasmic concentration of Ca^{2+} leads to Doux activation. It is, however, presently unknown how Ca^{2+} binding to the motifs affects their interaction with the C-terminal FNR-related domain. The activity of Duox1 and Duox2 is also regulated differentially via specific direct phosphorylation by PKA and PKC, respectively [27]. The Duox oxidases also rely on the Duox maturation factors DuoxA1 and/or DuoxA2 to reconstitute active Duox on the plasma membrane: congenital hypothyroidism is caused by genetic defect in the gene encoding DuoxA2 as well as that in the Duox2 gene [5, 6]. In the formation of an active Duox complex, DuoxA1 and DuoxA2 appear to function in a similar manner as $p22^{phox}$ by stably interacting with the partner oxidases on the plasma membrane [10, 11]. However a Duox regulatory protein that directly interacts with DuoxA1 and/or DuoxA2, similar to the $p22^{phox}$-binding proteins $p47^{phox}$ and Noxo1, has not been identified.

Nox Family Oxidases in Digestive Diseases

Two members of the Nox family, *i.e.*, Nox1 and Duox2, are highly expressed in the gastrointestinal tract [6, 7, 28]. In humans, Nox1 is abundant in epithelial cells of the colon but not in the stomach. On the other hand, its guinea pig homologue is highly expressed in gastric epithelial cells; the expression is strongly induced by bacterial lipopolysaccharide, suggestive of a role for Nox1 in host defense [28]. Nox1 as well as its partner proteins ($p22^{phox}$, Noxo1, and Noxa1) is significantly expressed in human gastric adenocarcinomas, although neither Nox1 nor its activating protein Noxo1 is present in the normal stomach. By contrast, in colon tumors, Nox1 is abundant in well-differentiated adnocarcinomas but scarcely expressed in poorly differentiated adenocarcinomas [28]. Future studies are needed to clarify the role of Nox1 in neoplastic transformation of gastrointestinal epithelial cells. The dual oxidase Duox2 is abundant in epithelial cells of the distal digestive tract, especially the colon and rectum; its significance remains to be elucidated.

The phagocyte oxidase Nox2 produces ROS at infectious sites, and ROS have been conventionally considered harmful mediators of tissue damage and inflammation. However, recent findings indicate that ROS produced by Nox2 also have an anti-inflammatory effect and thus prevent autoimmune disease [29–33]. Patients with CGD suffer from not only severe infections but also non-infectious inflammatory diseases such as granulomatous colitis [29–31], which have several properties common to Crohn's disease [30]; similarly, model animals for CGD exhibit a hyper-inflammatory phenotype [32, 33]. Thus ROS production through Nox2 as well as other Nox family oxidases could be a possible target for the inflammatory bowel diseases including Crohn's disease. The possibility should be tested in future studies.

Acknowledgements

This work was supported in part by Grants-in-Aid for Scientific Research and Targeted Proteins Research Program (TPRP) from the Ministry of Education, Culture, Sports, Science and Technology (MEXT) and by Japan Foundation for Applied Enzymology, Japan.

References

1 Lambeth JD: NOX enzymes and the biology of reactive oxygen. Nat Rev Immunol 2004;4:181–189.
2 Sumimoto H, Miyano K, Takeya R: Molecular composition and regulation of the Nox family NAD(P)H oxidases. Biochem Biophys Res Commun 2005;338: 677–686.
3 Dagher MC, Pick E: Opening the black box: lessons from cell-free systems on the phagocyte NADPH-oxidase. Biochimie 2007;89:1123–1132.
4 Bedard K, Krause K-H: The NOX family of ROS-generating NADPH oxidases: physiology and pathophysiology. Physiol Rev 2007;87:245–313.
5 Sumimoto H: Structure, regulation and evolution of Nox-family NADPH oxidases that produce reactive oxygen species. FEBS J 2008;275:3249–3277.
6 Leto TL, Morand S, Hurt D, Ueyama T: Targeting and regulation of reactive oxygen species generation by Nox family NADPH oxidases. Antioxid Redox Signal 2009;11:2607–2619.
7 Allaoui A, Botteaux A, Dumont JE, Hoste C, De Deken X: Dual oxidases and hydrogen peroxide in a complex dialogue between host mucosae and bacteria. Trends Mol Med 2009;15:571–579.
8 Brown DI, Griendling KK: Nox proteins in signal transduction. Free Radic Biol Med 2009;47:1239–1253.
9 Meitzler JL, Ortiz de Montellano PR: *Caenorhabditis elegans* and human dual oxidase 1 (DUOX1) 'peroxidase' domains: insights into heme binding and catalytic activity. J Biol Chem 2009;284:18634–18643.
10 Morand S, Ueyama T, Tsujibe S, Saito N, Korzeniowska A, Leto TL: Duox maturation factors form cell surface complexes with Duox affecting the specificity of reactive oxygen species generation. FASEB J 2009;23:1205–1218.
11 Luxen S, Noack D, Frausto M, Davanture S, Torbett BE, Knaus UG: Heterodimerization controls localization of Duox-DuoxA NADPH oxidases in airway cells. J Cell Sci 2009;122:1238–1247.
12 Matute JD, Arias AA, Wright NAM, Wrobel I, Waterhouse CCM, Li XJ, Marchal CC, Stull ND, Lewis DB, Steele M, Kellner JD, Yu W, Meroueh SO, Nauseef WM, Dinauer MC: A new genetic subgroup of chronic granulomatous disease with autosomal recessive mutations p40^{phox} and selective defects in neutrophil NADPH oxidase activity. Blood 2009;114:3309–3315.
13 Taura M, Miyano K, Minakami R, Kamakura S, Takeya R, Sumimoto H: A region N-terminal to the tandem SH3 domain of p47^{phox} plays a crucial role in the activation of the phagocyte NADPH oxidase. Biochem J 2009;419:329–338.
14 Yuzawa S, Miyano K, Honbou K, Inagaki F, Sumimoto H: The domain organization of p67^{phox}, a protein required for activation of the superoxide-producing NADPH oxidase in phagocytes. J Innate Immun 2009;1:543–555.
15 Sumimoto H, Kamakura S, Ito T: Structure and function of the PB1 domain, a protein interaction module conserved in animals, fungi, amoebas, and plants. Sci STKE 2007;2007:re6.
16 Miyano K, Koga H, Minakami R, Sumimoto H: The insert region of the Rac GTPases is dispensable for activation of superoxide-producing NADPH oxidases. Biochem J 2009;422:373–382.
17 Maehara Y, Miyano K, Sumimoto H: Role for the first SH3 domain of p67^{phox} in activation of superoxide-producing NADPH oxidases. Biochem Biophys Res Commun 2009;379:589–593.
18 Lassègue B, Griendling KK: NADPH oxidases: functions and pathologies in the vasculature. Arterioscler Thromb Vasc Biol 2010;30:653–661.
19 Nakano Y, Longo-Guess CM, Bergstrom DE, Nauseef WM, Jones SM, Bánfi B: Mutation of the *Cyba* gene encoding p22^{phox} causes vestibular and immune defects in mice. J Clin Invest 2008;118: 1176–1185.
20 Nagamani SC, Erez A, Shen J, Li C, Roeder E, Cox S, Karaviti L, Pearson M, Kang, SH, Sahoo T, Lalani, SR, Stankiewicz P, Sutton VR, Cheung SW: Clinical spectrum associated with recurrent genomic rearrangements in chromosome 17q12. Eur J Hum Genet 2010;18:278–284.

21 von Löhneysen K, Noack D, Jesaitis AJ, Dinauer MC, Knaus UG: Mutational analysis reveals distinct features of the Nox4-p22^{phox} complex. J Biol Chem 2008;283:35273–35282.
22 Helmcke I, Heumüller S, Tikkanen R, Schröder K, Brandes RP: Identification of structural elements in Nox1 and Nox4 controlling localization and activity. Antioxid Redox Signal 2009;11:1279–1287.
23 von Löhneysen K, Noack D, Wood MR, Friedman JS, Knaus UG: Structural insights into Nox4 and Nox2:motifs involved in function and cellular localization. Mol Cell Biol 2010;30:961–975.
24 Nisimoto Y, Jackson HM, Ogawa H, Kawahara T, Lambeth JD: Constitutive NADPH-dependent electron transferase activity of the Nox4 dehydrogenase domain. Biochemistry 2010;49:2433–2442.
25 Ago T, Kuroda J, Pain J, Fu C, Li H, Sadoshima J: Upregulation of Nox4 by hypertrophic stimuli promotes apoptosis and mitochondrial dysfunction in cardiac myocytes. Circ Res 2010;106:1253–1264.
26 Tirone F, Radu L, Craescu CT, Cox JA: Identification of the binding site for the regulatory calcium-binding domain in the catalytic domain of NOX5. Biochemistry 2010;49:761–771.
27 Rigutto S, Hoste C, Grasberger H, Milenkovic M, Communi D, Dumont JE, Corvilain B, Miot F, De Deken X: Activation of dual oxidases Duox1 and Duox2:differential regulation mediated by cAMP-dependent protein kinase and protein kinase C-dependent phosphorylation. J Biol Chem 2009;284:6725–6734.
28 Rokutan K, Kawahara T, Kuwano Y, Tominaga K, Nishida K, Teshima-Kondo S: Nox enzymes and oxidative stress in the immunopathology of the gastrointestinal tract. Semin Immunopathol 2008;30:315–327.
29 Marciano BE, Rosenzweig SD, Kleiner DE, Anderson VL, Darnell DN, Anaya-O'Brien S, Hilligoss DM, Malech HL, Gallin JI, Holland SM: Gastrointestinal involvement in chronic granulomatous disease. Pediatrics 2004;114:462–468.
30 Marks DJ, Miyagi K, Rahman FZ, Novelli M, Bloom SL, Segal AW: Inflammatory bowel disease in CGD reproduces the clinicopathological features of Crohn's disease. Am J Gastroenterol 2009;104:117–124.
31 Schäppi MG, Jaquet V, Belli DC, Krause K-H: Hyperinflammation in chronic granulomatous disease and anti-inflammatory role of the phagocyte NADPH oxidase. Semin Immunopathol 2008;30:255–271.
32 Romani L, Fallarino F, De Luca A, Montagnoli C, D'Angelo C, Zelante T, Vacca C, Bistoni F, Fioretti MC, Grohmann U, Segal BH, Puccetti P: Defective tryptophan catabolism underlies inflammation in mouse chronic granulomatous disease. Nature 2008;451:211–215.
33 Hultqvist M, Olsson LM, Gelderman KA, Holmdahl R: The protective role of ROS in autoimmune disease. Trends Immunol 2009;30:201–208.

Hideki Sumimoto
Department of Biochemistry, Kyushu University
Graduate School of Medical Sciences
3–1–1 Maidashi, Higashi-ku
Fukuoka 812–8582 (Japan)
Tel. +81 92 642 6096, Fax +81 92 642 6103, E-Mail hsumi@med.kyushu-u.ac.jp

Naito Y, Suematsu M, Yoshikawa T (eds): Free Radical Biology in Digestive Diseases.
Front Gastrointest Res. Basel, Karger, 2011, vol 29, pp 35–54

Neutrophil-Dependent Oxidative Stress in Inflammatory Gastrointestinal Diseases

Yuji Naito · Toshikazu Yoshikawa

Molecular Gastroenterology and Hepatology, Kyoto Prefectural University of Medicine, Kyoto, Japan

Abstract

Gastrointestinal inflammation is a highly complex biochemical protective response to cellular/tissue injury. When this process occurs in an uncontrolled manner, the result is excessive tissue damage that results chronic inflammation and destruction of normal tissue, and may associate with carcinogenesis. Current evidence suggests that neutrophil-dependent oxidative stress plays a crucial role in the pathogenesis of gastrointestinal inflammation associated with *Helicobacter pylori* infection, nonsteroidal anti-inflammatory drug ingestion, ischemic condition and inflammatory bowel disease. It has been shown that the neutrophil-vascular endothelial cell interaction is regulated by various cell adhesion molecules, and that this interaction is directly or indirectly modified by many factors, including bacterial products, chemokines and gaseous mediators. Activated neutrophils can produce reactive oxygen/nitrogen species, myeloperoxidase, and elastases. This review describes the potential role of neutrophils and neutrophil-dependent oxidative stress in gastrointestinal disease.

Although it is clear that genetic, environmental, and immunological factors affect the pathophysiology of the gastrointestinal diseases, recent studies have demonstrated that oxidative stress induced by the infiltration and activation of neutrophils plays a crucial role in the pathophysiology of inflammation of the gastrointestinal tract. Many experimental studies using neutrophil-deficient animals have clearly indicated the important role of neutrophils in the pathogenesis of the gastrointestinal inflammation, including *Helicobacter pylori*-induced gastritis, gastrointestinal inflammation induced by nonsteroidal anti-inflammatory drugs (NSAIDs) ingestion, and inflammatory bowel disease. The sequence of events in the infiltration and the activation of neutrophils have also been clarified, and some candidates have been proposed as a therapeutic target for these diseases during the last decade. More importantly in the clinical field, several studies have showed that granulocyte adsorptive apheresis therapy could induce the remission stage especially in patients with ulcerative colitis

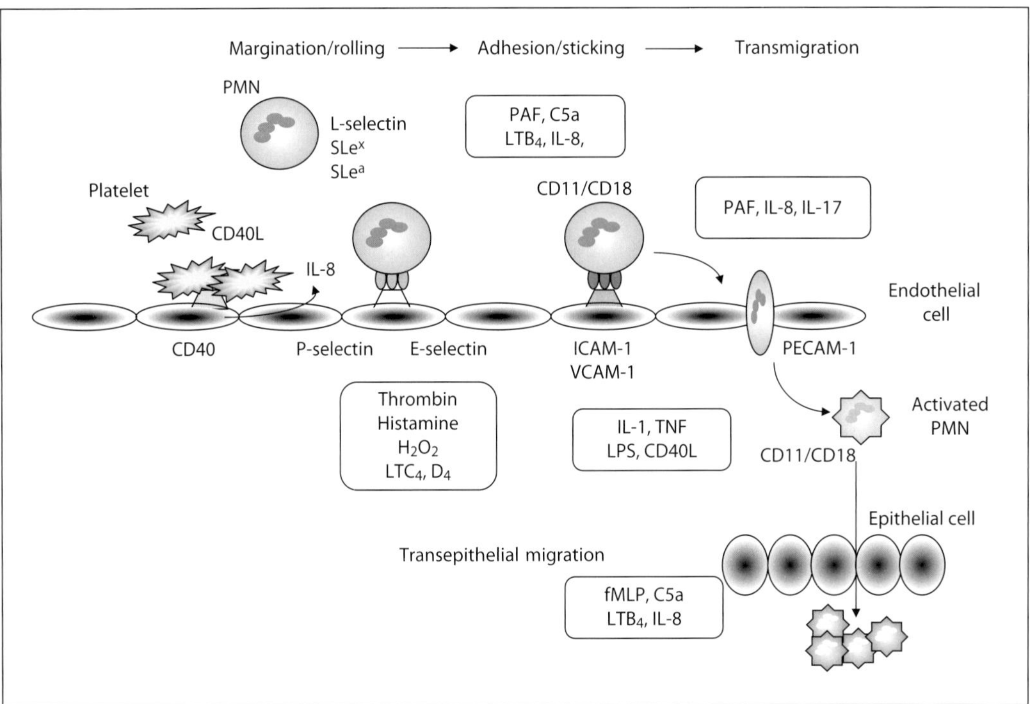

Fig. 1. Neutrophil-endothelial interaction mediated by adhesion molecules and chemical mediators. PAF = Platelet activating factor; PECAM-1 = platelet endothelial adhesion molecule 1; fMLP = N-formyl methionyl phenylalanine; LTC_4 = leukotriene C_4.

and Crohn's disease. Here, we review recent advances in molecular mediators of the infiltration and activation of neutrophils in gastrointestinal inflammatory disease.

Molecular Mediators in Neutrophil Infiltration

Leukocyte and Vascular Endothelial Cell Interaction

The sequence of events in the extravasation of neutrophils from the vascular lumen to the extravascular space is divided into (1) margination and rolling, (2) adhesion and transmigration, and (3) migration in interstitial mucosal tissues towards chemotactic stimulants (fig. 1), which are regulated by the interaction of adhesion molecules located on the surface of neutrophils and endothelial cells [1–3]. With stimuli, such as various cytokines and inflammatory mediators, neutrophils roll slowly on endothelial cells through interactions between L-selectin and carbohydrate antigen on neutrophils, and P- and E-selectin on endothelial cells. P-selectin is present in the Weibel-Plade bodies of endothelial cells and in the α-granules of platelets. Stimulation with histamine, thrombin, or reactive oxygen species (ROS) leads to rapid degranulation and translocation of P-selectin to the cell surface within a few minutes [4]. Eventually, the neutrophils adhere strongly to endothelial cells via CD11/CD18 glycoproteins and endothelial adhesion molecules of immunoglobulin superfamily, including the intercellular adhesion molecule 1 (ICAM-1) and the vascular cell adhesion molecule 1 (VCAM-1). ICAM-1 is expressed by activated and nonactivated endothelial cells and by lymphocytes, monocytes, and epithelial cells including gastric/intestinal epithelial cells. Its expression is stimulated by molecules such as interleukin (IL)-1, IL-17, tumor necrosis factor-α (TNF-α), interferon-γ (IFN-γ), and lipopolysaccharide (LPS). ICAM-1 binds β_2-integrins lymphocyte function-associated antigen-1 (LFA-1; CD11a/CD18) and Mac-1 (CD11b/CD18), which are stored in the intracellular granules of leukocytes. VCAM-1 is found in a membrane-bound form on various cells, including endothelial cells and monocytes/macrophages and also exists in a soluble form. The expression of VCAM-1 is stimulated by TNF-α, IL-1β, IL-4, IFN-γ, and LPS. After firmly binding to the endothelial surface (primarily CD11/CD18 binding to ICAM-1), the leukocytes transmigrate between cells along the intercellular junction. Platelet endothelial cell adhesion molecule 1, a cell-cell adhesion molecule, is a likely candidate for mediating this process. After passing the endothelial junctions, leukocytes are able to cross the basement membrane by focally degrading it with secreted collagenases. Due to the crucial role played by these adhesion molecules in the pathogenesis of gastrointestinal inflammatory diseases, targeting of these molecules has recently been proposed as a new direction for the development of anti-inflammatory strategies for these diseases. However, systemic treatment with alicaforsen (ISIS 2302), an antisense to ICAM-1 for Crohn's disease has not revealed significant effect, which was demonstrated by a randomized, double-masked, placebo-controlled study [5].

Cytokines and Mediators

After extravasation, neutrophils emigrate towards the site of gastrointestinal mucosal injury along a chemical gradient of chemotaxis. In the gastrointestinal inflammation, exogenous and endogenous substances are able to act as chemotactic agents for neutrophils including (1) cytokines especially IL-8, (2) soluble bacterial products, particularly peptides with N-formyl-methionine termini, (3) components of complement system, and (4) products of the lipoxygenase pathway of arachidonic acid metabolism, particularly leukotriene B_4 (LTB_4). IL-8, a member of the CXC chemokine family, is an important activator and chemoattractant for neutrophils, and has been implicated in the pathogenesis of a variety of inflammatory disease, including reflux esophagitis [6, 7], *H. pylori*-induced gastritis [8–11], NSAID-induced gastrointestinal injuries [12, 13], ischemic intestinal injury [14–16], and inflammatory bowel disease [17–19]. Especially, it has been indicated that a strong correlation exists between mucosal IL-8 contents and the level of polymorphonuclear cell infiltration in *H. pylori*-associated antral inflammation [20].

Current data suggest that Toll-like receptors (TLRs), which recognize specific pathogen-associated molecular patterns, are differentially expressed on both leukocytes and mucosal epithelial cells while serving to modulate leukocyte-epithelial interactions. Exposure of epithelial TLRs to microbial ligands has been shown to result in transcriptional upregulation of inflammatory mediators, whereas ligation of leukocyte TLRs modulates specific antimicrobial responses. Recently, Watanabe et al. [21] have reported that TLR4-deficient and MyD88-deficient mice are resistant to the intestinal damage induced by indomethacin. It has also been reported that treatment with CRX-526, a synthetic TLR4 antagonist, inhibits the development of moderate-to-severe disease in two mouse models of colonic inflammation: the dextran sodium sulfate model and multidrug resistance gene 1α-deficient mice [22]. These data indicate the significant role of TLR4-dependent pathway in the pathogenesis of the intestinal inflammation. A better understanding of these events will hopefully provide new insights into the mechanisms of epithelial responses to microorganisms and ideas for therapies aimed at inhibiting the deleterious consequences of mucosal inflammation in gastrointestinal disease.

As one of effectors of mucosal inflammation, a new lineage of effector $CD4^+$ T cells characterized by production of IL-17, the T-helper (Th)-17 lineage, was recently described [23]. It has been shown that IL-17-expresing T cells are increased in the inflamed mucosa of inflammatory bowel disease patients, and that serum IL-17 concentrations are elevated in patients with active inflammatory bowel disease [24]. Recent studies have demonstrated that IL-23 and transforming growth factor-β can induce the differentiation of naïve T cells into Th-17 cells, and that this differentiation is most efficiently induced when Th1 and Th2 effector functions are simultaneously inhibited [25–27]. Interestingly, IL-17 appears to stimulate predominantly the production of cytokines that either specifically attract neutrophils to the site of inflammation (IL-8, GROα, GCP-2) or stimulate granulopoiesis in bone marrow (IL-6, G-CSF, GM-CSF) [28]. While IL-17 itself does not affect neutrophil chemotaxis, the supernatants from IL-17-treated fibroblast, epithelial and endothelial cells stimulate neutrophil migration. IL-17 appears to be an important mediator of inflammation, especially in neutrophil-dominated response to bacterial challenge, and may enhance faster and more effective recruitment of neutrophil as an important element of host defense.

Role of Platelets

Recently, a great deal of attention has been given to the role of platelets in gastrointestinal pathophysiology [29]. Patients with inflammatory bowel disease exhibit an increased expression of CD40L on platelets and increased plasma levels of soluble

CD40L, which is largely derived from the activate CD40L$^+$ platelets [30–32]. The most prominent activities mediated by the platelet CD40/CD40 ligand pathway include inflammatory, immunoregulatory, and hemostatic functions, all of which contribute to the newly expanded view of platelets as key biological mediators involved in inflammatory bowel disease [33]. Through the CD40-dependent pathway, inflammation-derived platelets stimulate intestinal microvascular endothelial cells to enhance the ICAM-1/VCAM-1 expression and to produce IL-8, setting in motion endothelial cell signaling machinery through the mitogen-activated protein kinase cascade, and promoting phosphorylation of p38. Using an experimental colitis model, Vowinkel et al. [34] have recently shown that platelet-leukocyte interactions are mediated by CD40L, as revealed by the significant reduction in circulating platelet-leukocyte aggregates if CD40L-deficient mice are used. These data suggest that modulation of leukocyte and platelet recruitment by activated CD40-positive endothelial cells in intestinal venules may represent a major action of CD40-CD40L signaling pathway.

In addition to CD40-CD40L, it has been reported that platelets can interact with neutrophils through at least four separate receptor-ligand pairs, including (platelet-PMN) GPIIb/IIIa–Mac-1, GPIIba–Mac-1, ICAM-2–LFA-1, and P-selectin–P-selectin glycoprotein ligand-1 (P-selectin–PSGL-1) [35]. Further study should be performed to elucidate the exact molecular details involved in platelet-neutrophil interactions.

Neutrophil and Epithelium Interaction

Neutrophil retention within mucosal layer and crypt of the gastrointestinal tract is believed to contribute to the development of chronic active gastritis and crypt abscess in ulcerative colitis. Therefore, preventing neutrophil transepithelial migration may be one means of preventing excessive inflammation in these patients. Hofman et al. [36] used confluent polarized monolayers of the human intestinal cell line T84 grown on permeable filters to analyze the epithelial neutrophil response induced by *H. pylori*, and showed that the vacuolating cytotoxin VacA is not involved in neutrophil transepithelial migration and that the *cag* pathogenicity island, with a pivotal role for the *cagE* gene, provokes a transcellular signal across T84 monolayers, inducing a sub-epithelial neutrophil response. It has also been shown that neutrophils migrate across intestinal epithelium using CD11b/CD18-dependent and -independent mechanisms [37–39]. For example, migration in response to the chemoattractant n-formyl methionyl phenylalanine is blocked with anti-CD11b or anti-CD18 antibodies [39]. In contrast, a substantial amount of neutrophil migration to chemoattractants complement C5a, IL-8 and LTB_4 occurs despite the presence of anti-CD11b or anti-CD18 antibody, with C5a being the most potent inducer of CD11b/CD18-independent migration [37, 38]. Additional investigation is required to identify the ligands supporting CD11b/CD18-independent migration.

Recent data suggest the important role of transmigrated neutrophils and platelets in the epithelial function. Although initial studies revealed that platelets, per se, lack the ability for significant migration capacity, they appear to follow neutrophils across the epithelium in the presence of neutrophils. During active inflammation, platelets are caught in the flow of neutrophil transmigration, resulting in platelet translocation across the apical site of mucosal epithelial cells [40]. Neutrophils and platelet-derived ATP are selectively metabolized to adenosine by a two-step enzymatic reaction involving ecto-apyrase and ecto-nucleotidase (CD73). Adenosine binding to apical adenosine A2B receptors results in activation of electrogenic Cl^- secretion from intestinal epithelial cells and the paracellular movement of water. Such a platelet/neutrophil-epithelial crosstalk pathway may serve as a defensive response by which mucosal surfaces are flushed from bacteria and bacterial products under inflammatory conditions [40].

Role of Macrophages/Monocytes

Yamamoto et al. [41] have clearly shown a key role of the Ca^{2+}-permeable channel melastatin-like transient receptor potential 2 (TRPM2) in hydrogen peroxide (H_2O_2)-induced IL-8 chemokine production in monocytes that is of major physiological consequence in inflammation. They have demonstrated that H_2O_2 activates TRPM2 and amplifies downstream Ras and Erk signaling via Pyk2, leading to nuclear translocation of NF-κB and production of IL-8 in the human monocytic cell line U937; they also showed in *Trpm2*-knockout mice that TRPM2 controls CXCL2, similar to IL-8 in humans, production in monocytes, which induces neutrophil migration and exacerbates dextran sulfate sodium-induced ulcerative colitis. Inflammation parameters such as neutrophil infiltration and ulceration in the colon are also reduced in dextran sulfate sodium-treated *Trpm2*-knockout mice. These studies suggest a key role of IL-8 produced by TRPM2 in monocytes and macrophages in the exacerbation of human inflammatory bowel disease. This raises the prospect that suppression of IL-8 production by inhibition of TRPM2 might be an effective way to reduce pathological severity in ulcerative colitis and many other inflammatory diseases related to ROS production [41].

Neutrophil Activation in Gastrointestinal Diseases

Neutrophil Activation in H. pylori Infection

H. pylori is a noninvasive, non-spore-forming, G-shaped Gram-negative rod bacteria measuring approximately 3.5 × 0.5 μm. The mechanism by which gastritis develops remains unclear but may relate to the combined influence of bacterial enzymes and toxins and to noxious chemicals by the recruited neutrophils. The initial response to *H. pylori* infection in human appears to have a marked neutrophil infiltration, and this response is probably induced both directly by bacterial factors and indirectly via the induction of chemokines involved in the inflammatory cascade. Yoshida et al. [42] have demonstrated that CD11b/CD18 (Mac-1) is expressed on the cell membrane of neutrophils, and they adhere strongly to human endothelial cells when neutrophils are activated by a water extract of *H. pylori*. Because this adhesion is inhibited by a monoclonal antibody directed against CD11b, CD18, or ICAM-1, it would seem that the adhesion takes place through interaction of CD11b/CD18 on neutrophils and ICAM-1 on endothelial cells. Substances that cause neutrophils to express adhesion molecules did not contain previously known neutrophil-activating substances, such as platelet-activating factor and LTB_4.

A major proinflammatory factor produced by *H. pylori* is *H. pylori* neutrophil-activating protein (HP-NAP) [43]. It was originally defined as polymorphonuclear leukocyte (PMN)-activating protein because it stimulates PMN to produce ROS [44]. In addition, it has been shown that HP-NAP (a) is a major antigen in human immune response to *H. pylori*; (b) is capable, after oral immunization, of conferring protection on mice against a subsequent challenge with *H. pylori*; (c) activates the NADPH oxidase of human leukocytes via a set of signaling events involving a pertussis toxin-sensitive G protein, an elevation of cytosolic $[Ca^{2+}]$, an Src family of tyrosine kinases, and wortmannin-sensitive PI3 kinase [45, 46], and (d) stimulates monocytes/macrophages to produce TNF-α and IL-8. TNF-α is a pleiotropic cytokine able to stimulate adhesivity of endothelial cells by upregulation adhesion molecules, such as VCAM-1 and ICAM-1. Moreover, TNF-α can induce activation of integrins on neutrophils, directly or by stimulating the secretion of IL-8 from the endothelium. More recently, Polenghi et al. [47] have demonstrated that HP-NAP not only recruits leukocytes from the vascular lumen, but also stimulates them to produce messengers that may contribute to the maintenance of chronic active inflammation associated with the *H. pylori* infection using intravital microscopy analysis of rat mesenteric venules.

IL-8 production from gastric epithelial cells stimulated with *H. pylori* may contribute to the activation of neutrophils in *H. pylori*-associated chronic gastritis. In vitro studies have shown that the expression of IL-8 in gastric epithelial cells, such as MKN45 cells, MKN-28 cells, and AGS cells, is upregulated by adhesion of *H. pylori* to gastric epithelial cells [48–50] or the extract or surface proteins of this bacterium [11, 51]. We also found that the water extract of *H. pylori* activates gastric epithelial cells MKN-45, and causes the expression of IL-8 mRNA and protein [11]. This indicates that live bacteria are not essential to stimulate IL-8 production, since water extract containing only soluble factors of *H. pylori* also stimulated IL-8 production in gastric epithelial cells.

Neutrophil Activation in NSAID-Induced Gastrointestinal Injury

Recent reports have hypothesized that neutrophil-mediated inflammation is involved in the development of NSAID-induced gastrointestinal mucosal injuries. Five lines of evidence support this hypothesis: (1) neutrophil depletion by intraperitoneal injection of antineutrophil antibody significantly attenuates these injuries, (2) intravital microscopic approaches applied to the mesenteric circulation have shown that indomethacin or aspirin promotes leukocyte adherence and emigration in postcapillary venules, (3) immunoneutralization of CD11/CD18 adherence complex on neutrophils attenuates these injuries, (4) these drugs directly promote neutrophil adherence to endothelium via CD11b/CD18-dependent interactions with ICAM-1, and (5) adhesion molecules CD11a and ICAM-1 are expressed in normal gastric mucosa, and the number of ICAM-1-stained blood vessels rapidly increase after indomethacin treatment. The mechanisms responsible for the upregulation of ICAM-1 after indomethacin/aspirin administration are unclear. TNF-α is one of the candidates for a mediator involved in NSAID-induced leukocyte adherence. TNF-α is a proinflammatory cytokine and has recently been shown to be a crucial mediator of NSAID-induced gastric mucosal injury [52, 53]. The addition of a TNF-α processing enzyme inhibitor has been shown to prevent aspirin-induced TNF-α release and protect against gastric mucosal injury in rats [54]. TNF-α augments neutrophil-derived superoxide generation [55] and upregulates the expression of adhesion molecules on neutrophils and endothelial cells [56], leading to neutrophil accumulation and oxygen radical-mediated tissue damage. Our recent study has shown the reduced intestinal inflammation induced by indomethacin in TNF-α-deficient mice. These results indicate that a crucial initial event in the pathogenesis of NSAID-induced injury appears to be adherence of neutrophils to microvascular endothelium, and that drugs targeting TNF-α could be a novel strategy to treat NSAID-induced gastrointestinal injuries.

It has been speculated that NSAID-induced gastrointestinal lesions result from a deficiency of mucosal prostaglandins by the inhibition of cyclooxygenase. In addition, increased formation of leukotrienes, the other major products that are derived from arachidonic acid, may be involved in the pathogenesis of these injuries. Asako et al. [57] have demonstrated that both the reduction in leukocyte rolling velocity and increased leukocyte adherence induced by indomethacin are completely prevented by pretreatment with either a leukotriene synthesis inhibitor (L 663,536) or a LTB_4 receptor antagonist (SC 42930), while an LTD_4 receptor antagonist had no effect. In addition, indomethacin resulted in an 80% increase in LTB_4 release into mesenteric bathing solution. These observations suggest that inhibition of the cyclooxygenase pathway with indomethacin leads to an increased flux of arachidonate through the lipoxygenase pathway with a concomitant increase in leukotriene production. Miura et al. [58] have also demonstrated that AA-861, a selective inhibitor of 5-lipoxygenase, attenuates the microcirculatory disturbance and neutrophil accumulation in the intestinal mucosa after indomethacin treatment. From these observations, it is considered that leukocyte accumulation and 5-lipoxygenase products, possibly LTB_4, are involved in the development of NSAID-induced intestinal ulcer.

Neutrophil Activation in Ulcerative Colitis

In patients with ulcerative colitis, the circulating levels of neutrophils are found to be up to three times higher than the levels in healthy controls [59]. Morphological and functional evidence of activation of circulating neutrophils has been reported in patients with inflammatory bowel disease. In 1991, McCarthy et al. [60] found the increase in the number of polarized neutrophils, determined by quantitative light microscope examination, in patients with active stage of ulcerative colitis compared to those with quiescent colitis or normal subjects. Functionally, neutrophils obtained from patients with ulcerative colitis have a significantly higher response than those from controls following phorbol myristate acetate, formyl-methionyl-leucyl-phenylalanine, and zymosan administration [61].

Anezaki et al. [17] have investigated the correlation between IL-8, and myeloperoxidase (MPO) or luminal-dependent chemiluminescence in inflamed mucosa of ulcerative colitis. They have found that luminol-dependent chemiluminescence of biopsy specimens in active ulcerative colitis is markedly increased compared to those in inactive ulcerative colitis and controls, and that the levels of IL-8 are closely correlated with the intensity of chemiluminescence or MPO levels. In ulcerative colitis, IL-8 mRNA was found mainly in macrophages, and also in neutrophils and colonic epithelial cells [19]. In addition, increased production of IL-8 peptides and expression of IL-8 mRNA is observed in the inflamed mucosa of patients with ulcerative colitis [62]. Since IL-8 is not only a chemoattractive substance but also a neutrophil-activating substance to release oxygen radicals from neutrophils, the data by Anezaki et al. [17] suggest that most of the infiltrating neutrophils in the colonic mucosa of ulcerative colitis are activated.

Human neutrophil lipocalin (HNL) may be a more sensitive marker for neutrophil activation in the colonic mucosa compared to MPO, because primary granules including MPO are not unique to the neutrophil granulocyte. HNL is a secondary granule protein unique to the neutrophil. Carlson et al. [63] have demonstrated that HNL levels in colorectal perfusion fluids increase in ulcerative colitis and proctitis, and suggest that HNL may serve as a specific marker of intestinal neutrophil activation in ulcerative colitis.

The granulocyte and monocyte adsorptive apheresis (GMCAP; Adacolumn) is an example of a medical device that can selectively remove activated granulocytes and monocytes/macrophages together with small populations of lymphocytes. Although the aim of GMCAP has been to remove excess and activated granulocytes and monocytes from the circulation, recent reports have demonstrated the anti-inflammatory and immunomodulatory actions of this leukocytapheresis [64]. The efficacy of GMCAP was firstly evaluated using clinical activity index or disease activity index scores in patients with ulcerative colitis refractory to conventional drug therapy [65]. In a prospective multicenter trial in Japan, 53 patients were treated with the standard GMCAP protocol (five apheresis sessions for 5 consecutive weeks) in combination with prednisolone. After the treatment, 21 and 37% of patients achieved remission and improvement, respectively, and the mean daily dose of prednisolone per patient was reduced from 24.4 mg at enrollment to 14.2 mg after GMCAP therapy. In addition to steroid-refractory ulcerative colitis patients, it has been shown that more than 85% of steroid-naive patients achieved remission by GMCAP therapy [66, 67]. In summary, clinical data obtained in Japan suggest that leukocytapheresis is an effective adjunct to therapy for ulcerative colitis to promote remission, taper conventional drug dosage, and potentially reduce the number of patients who require colectomy. In addition to ulcerative colitis, recent studies have investigated the efficacy of GMCAP therapy for Crohn's disease patients in Japan [68] and Scandinavia [69]. In the study from Japan, 21 patients with a Crohn's disease activity index of 200–399 received GMCAP as an adjunct to their ongoing medication. These results indicate that GMCAP could be effective for inducing remission and improving quality of life in patients with active Crohn's disease that is refractory to conventional therapy [68].

Neutrophil-Dependent Mucosal Injury

Lipid Peroxidation

In activated neutrophils, NADPH oxidase in cell membranes becomes activated, and an electron transfer takes place from NADPH in cells to oxygen inside and outside cells, and the oxygen that received electrons becomes superoxide radicals ($O_2^{-\cdot}$), which is rapidly converted to H_2O_2 by spontaneous dismutation or enzymatic superoxide dismutase, and hydroxyl radicals, which are formed nonenzymatically in the presence of Fe^{2+} as a secondary reaction [70, 71].

ROS are highly reactive. When they are generated close to cell membranes, possibly by gastrointestinal epithelial cells, they induce oxidative stress and oxidized membrane phospholipids (lipid peroxidation), which may continue in a form of a chain reaction. Biomembranes contain large amounts of polyunsaturated fatty acids (PUFAs) in their phospholipids. PUFAs contain two or more carbon-double bonds within their structure. This makes them susceptible to oxidative damage by free radical attack, which is a direct cell injury induced by oxidative stress. PUFAs of cell membranes are degraded by lipid peroxidation with subsequent disruption of membrane integrity, suggesting that lipid peroxidation mediated by oxygen radicals is an important cause of damage and destruction of cell membranes. We have shown in a model of the gastrointestinal inflammation that thiobarbituric-reactive substances, an index of lipid peroxidation, significantly accumulate in the mucosa, and that the treatment with superoxide dismutase led to an improvement in the inflammation of the gastrointestinal tract [72–74]. In addition, treatment with a synthetic vitamin E analogue inhibits the intestinal inflammation in mice [75]; this suggests that induction of lipid peroxidation associated with superoxide production is an early critical event in these experimental models of gastrointestinal inflammation.

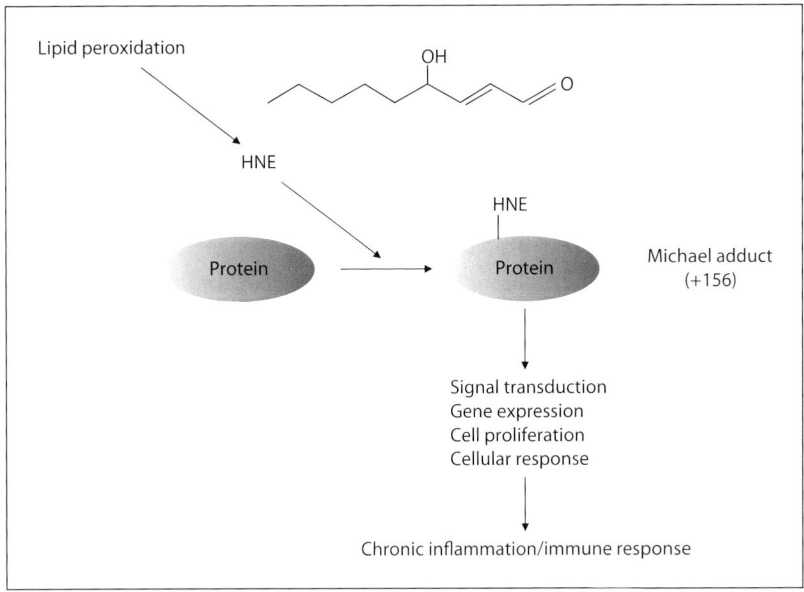

Fig. 2. HNE formation and its adducts as a second messenger.

In addition to direct products derived from activated neutrophils, the secondary products induced by oxidative stress may play a role in the development of gastrointestinal inflammation. Experimental and clinical evidence coming from different laboratories suggest that 4-hydroxy-2-nonenal (HNE; fig. 2), a product during lipid peroxidation, can act as bioactive molecules in either physiological or pathological conditions [76]. HNE can affect and modulate, at very low and nontoxic concentrations, several cell functions, including signal transduction, gene expression, cell proliferation, and more generally, the response of the target cells [76–78]. Nair et al. [79, 80] have demonstrated the accumulation of lipid peroxidation-derived DNA bases, which are generated by reaction of DNA with HNE, in the colon of patients with inflammatory bowel disease, and suggested that these HNE-DNA adducts may serve as potential lead markers for assessing progression of inflammatory cancer-prone disease. Recently, our preliminary study has also demonstrated HNE-modified proteins in colonic mucosa obtained from patients with ulcerative colitis using mono- and polyclonal antibodies against HNE. Among these HNE-modified proteins, Kimura et al. [81] have found the increase in HNE production and modification of IgA with HNE in the intestinal mucosa of the rats shortly after LPS injection, and showed that the HNE modification promotes IgA polymerization and secretion by causing oxidation of free sulfhydryl groups of IgA dimers. These data indicate the involvement of HNE, a product of lipid peroxidation, in the immune response of plasma cells in early intestinal inflammation.

The interaction of HNE with a variety of kinases variously involved in cell signaling associated with inflammation is now a matter of active investigation. In particular, findings with regard to the effect of HNE on different components of the protein kinase C family and the mitogen-activated protein kinase complex already provide reliable indications of a potential role of this aldehyde as a cell signal messenger. Recently, Biasi et al. [82] reported that c-Jun N-terminal kinase upregulation plays a key role in the proapoptotic interaction between transforming growth factor-β_1 and HNE in colon mucosa. Such a role appears further supported by the clear-cut evidence of upregulation of receptor tyrosine kinases and downregulation of the nuclear factor kappa B system, produced by HNE concentrations actually detectable in pathophysiology of intestinal inflammation. Using antibodies against HNE, we can indicate the distribution and the target molecules of this active molecule as a molecular fingerprint of lipid peroxidation. Recent advances in proteomics technology make it possible to determine specific targets modified by oxidative stress, including HNE. This modification of proteins may be a biomarker for evaluation of disease activity or therapeutic efficacy.

Myeloperoxidase-Dependent Oxidative Stress

The toxicity of H_2O_2 is enhanced by the activity of MPO. MPO is abundant in primary azurophilic granules of leukocytes including neutrophils and monocytes/macrophages, and secreted into phagolysosomal compartment following phagocyte activation. In combination with H_2O_2, MPO can oxidize the halides and the pseudohalide thiocyanate (SCN^-) to their corresponding hypohalous acids.

$$H_2O_2 + HX \xrightarrow{\text{MPO}} HOX + H_2O \; (X = Cl^-, Br^-, I^-, \text{or } SCN^-)$$

Owing to its high concentration in biological fluids, Cl^- is the major substrate for MPO; consequently, hypochlorous acid (HOCl) is a major oxidation product. HOCl can oxidize sulfhydryl groups, halogenate and oxygenate unsaturated lipids, and halogenate aromatic compound. MPO-derived chlorinating agents also generate secondary oxidants such as monochloramines (NH_2Cl), dichloramines, and amino acid-derived aldehydes. NH_2Cl, a product derived from the interaction of NH_3 (a product by *H. pylori*) and HOCl (a product by activated neutrophil), is reported to be exceptionally reactive and toxic because of its high lipophilic property and low molecular weight [83]. NH_2Cl is bactericidal and mutagenic, inhibits the hexose-monophosphate pathway in eucaryotic cells, oxidizes erythrocyte hemoglobin and glutathione, and stimulates rat colonic secretion [84, 85]. Suzuki et al. [86] demonstrated that *H. pylori*-activated neutrophils promote gastric mucosal cell injury and that NH_2Cl, at concentrations of $\geq$0.1 mM, plays a unique and important role in this process. We have demonstrated that NH_2Cl inhibits gastric mucosal cell growth and induces apoptosis in rat normal gastric mucosal cells [87]. Because NH_2Cl is a potent mediator in the oxidation of DNA and in the induction of apoptosis [88], it is generally considered that NH_2Cl is a target molecule in our understanding of the mechanism

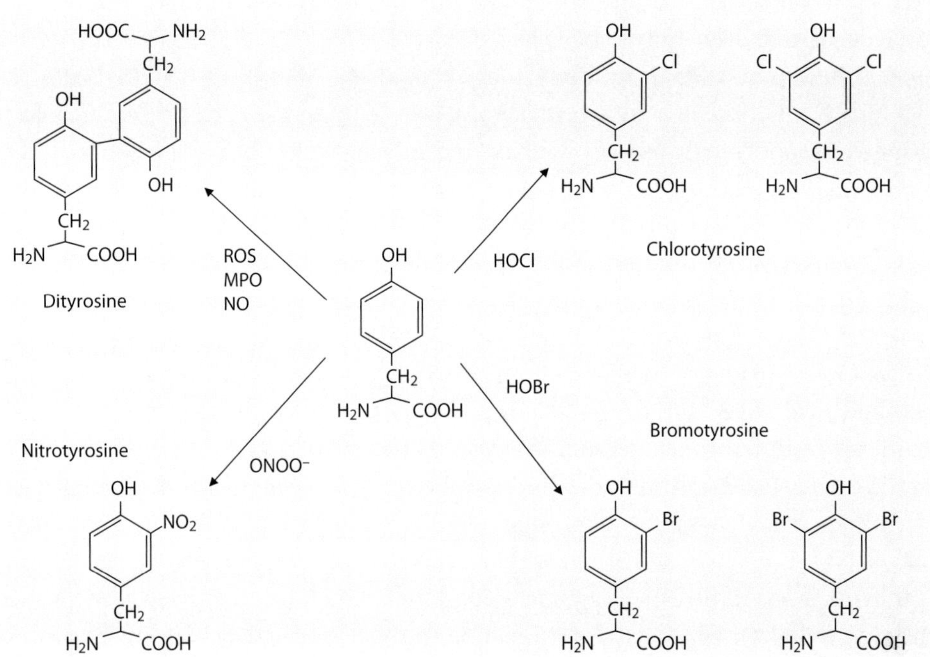

Fig. 3. Neutrophil-dependent modification of tyrosine residues of proteins.

of *H. pylori*-associated gastric carcinogenesis, or developing a novel chemopreventive agent.

Tyrosine as a Target of Oxidative/Nitrosative Stress

One of the targets for oxidative/nitrosative stress is the molecule 'tyrosine' in both its free and its bound forms. When it reacts with HOCl, tyrosine is converted into the 3-chlorotyrosine molecule (fig. 3). This molecule could also cause tissue damage or dysfunction in many inflammatory conditions. In addition, 3-chlorotyrosine is unique because it is heat stable and is not readily formed by artificial mechanism, which makes it an excellent marker, a 'molecular fingerprint', for MPO-induced oxidation [89]. Recent in vivo studies indicate that assessment of 3-chlorotyrosine protein adduct formation by immunohistochemistry could be a useful marker of neutrophil-induced cell injury in drug-induced liver injury [90], inflammatory neuronal degeneration [91], atherosclerosis [92] and cystic fibrosis [93].

In addition, it has been suggested that these reactive intermediates generated by phagocytes might damage nucleic acids in cells at the site of inflammation, and that these chlorinated (or brominated) nucleobases are endogenously generated in vivo

[94]. However, a detailed mechanism for the formation of halogenated nucleobases in vivo has not been fully established. Osawa's group [95, 96] has recently developed a new monoclonal antibody that recognizes 8-bromo-2′-dG (8-BrdG) and 8-chloro-2′-dG (8-CldG) to investigate the role of these molecules in inflammation. Using this monoclonal antibody, we have recently demonstrated that positive staining for 8-BrdG/8-CldG is observed in liver tissue from hepatocellular carcinoma patients, but not in liver tissue from human cirrhosis patients [95]. The use of these antibodies may give us new information about the role of neutrophil-derived hypohalous acids in the nuclear DNA damage of target tissues at sites of inflammation.

As one of the most important mediators of MPO, nitrotyrosine plays a key role in the process of oxidation seen on inflammatory conditions including *H. pylori* infection and ulcerative colitis. One important aspect of the formation of nitrotyrosine that must be noted is that it is not solely generated by peroxynitrite, a product of the interaction between nitric oxide (NO) and $O_2^{\cdot-}$. MPO can also independently oxidize nitrite (NO_2^-), a stable end product of NO metabolism, to form nitrogen dioxide (NO_2). NO_2 is also a reactive nitrogen species that in turn can nitrogenate tyrosine. A third pathway utilizes MPO-generated HOCl, which also oxidizes NO_2^- to nitryl chloride (NO_2Cl), which is also a reactive nitrogen species. Several recent investigations validate the importance of nitrotyrosine. There have been reports of the increases in nitrotyrosine-immunoreactive cells in experimental colitis induced by dextran sodium sulfate [97], dinitrobenzene sulfonic acid [98], and trinitrobenzene sulfonic acid [99] as well as in patients with inflammatory bowel disease [100–102]. Keshavarzian et al. [103] have reported that nitration of cytoskeletal protein actin is increased in the mucosa of patients with ulcerative colitis and Crohn's disease, regardless of type of activity, and suggested that oxidant-induced cytoskeletal disruption is required for tissue injury, mucosal disruption, and inflammatory bowel disease flare-up. Unexpectedly, the increased nitrotyrosine expression in ulcerative colitis is associated with neutrophilic MPO and not with NO synthase [100], which suggests that nitrotyrosine is formed in areas with a dense neutrophilic infiltrate via a peroxynitrite-independent oxidation pathway. We also demonstrated the presence of nitrotyrosine-modified proteins in the gastric mucosa of the patients with severe atrophic gastritis using a new anti-nitrotyrosine monoclonal antibody [101]. We hypothesize that identification of specific nitrated proteins would provide mechanistic information regarding nitrosative stress and insight into downstream functional consequences in *H. pylori*-infected gastric inflammation.

Elastase

Intracellular neutrophil elastase is a key effector molecule of the innate immune system, with potent antimicrobial activity against Gram-negative bacteria, spirochaetes, and fungi. In general, neutrophil elastase is capable of digesting virtually every type of matrix protein, including several types of collagen, fibronectin, proteoglycans, heparin, and cross-linked fibrin. To minimize tissue damage, serine protease inhibitors are secreted to neutralize aggressive proteases. Recently, Schmid et al. [104] have demonstrated that the induction of anti-protease Elafin and secretary protease inhibitor SLPI is attenuated in Crohn's disease, and have suggested that the balance between two important serine antiproteases and neutrophil elastase expression is drastically shifted toward the protease in Crohn's disease compared with ulcerative colitis.

Morohoshi et al. [105] also investigated the contribution of neutrophil elastase in a murine acute colitis model. They have shown that the neutrophil elastase enzyme activity is significantly elevated in both the plasma and colonic mucosal tissue in ulcerative colitis patients compared with healthy controls, and that ONO-5046, a neutrophil elastase-specific inhibitor exerts therapeutic effects in DSS-treated mice by significantly reducing weight loss and histological score. In addition to its protease activity contributing to tissue destruction, it has been reported that neutrophil elastase enhances the migration and adhesion of neutrophils [106]. ONO-5046 has already been clinically used for the treatment of acute respiratory distress syndrome in Japan, and no serious adverse effects have been reported. Therefore, ONO-5046 might actually have the potential to be a new therapeutic approach for patients with ulcerative colitis.

Conclusion

We have summarized the recent advances in the role of neutrophil-dependent oxidative stress in gastrointestinal inflammation. Although basic research has demonstrated that several antioxidants and inhibitors of neutrophil-derived enzymes inhibit the inflammatory responses in the gastrointestinal tract, further studies will be needed to clarify the precise mechanism of these agents. Clinically, the successful outcome of leukocytapheresis for inflammatory bowel disease clearly confirms the crucial role of neutrophils in the pathogenesis of this disease. To obtain more precise scientific evidence for the efficacy of antineutrophil/antioxidative therapies against gastrointestinal inflammation, oxidation-modified proteins as molecular fingerprint should be used for the evaluation of these therapies.

Acknowledgment

This work was supported by a Grant-in-Aid for Scientific Research (B) to T.Y. (No. 21390184) and Challenging Exploratory Research to Y.N. (No. 08101559) from Japan Society for the Promotion of Science, by a City Area Program to T.Y. and Y.N. from Ministry of Education, Culture, Sports, Science and Technology, Japan, and by a Adaptable and Seamless Technology Transfer Program through target-driven R&D to Y.N. from Japan Science and Technology Agency.

References

1 Kokura S, Wolf RE, Yoshikawa T, et al: Molecular mechanisms of neutrophil-endothelial cell adhesion induced by redox imbalance. Circ Res 1999;84:516–524.

2 Naito Y, Takagi T, Yoshikawa T: Neutrophil-dependent oxidative stress in ulcerative colitis. J Clin Biochem Nutr 2007;41:18–26.

3 Yoshida N, Yamaguchi T, Nakagawa S, et al: Role of P-selectin and intercellular adhesion molecule-1 in TNB-induced colitis in rats. Digestion 2001;1: 81–86.

4 Tedder TF, Steeber DA, Chen A, et al: The selectins: vascular adhesion molecules. FASEB J 1995;9:866–873.

5 Yacyshyn B, Chey WY, Wedel MK, et al: A randomized, double-masked, placebo-controlled study of alicaforsen, an antisense inhibitor of intercellular adhesion molecule 1, for the treatment of subjects with active Crohn's disease. Clin Gastroenterol Hepatol 2007;5:215–220.

6 Isomoto H, Saenko VA, Kanazawa Y, et al: Enhanced expression of interleukin-8 and activation of nuclear factor kappa-B in endoscopy-negative gastroesophageal reflux disease. Am J Gastroenterol 2004;99: 589–597.

7 Yoshida N, Uchiyama K, Kuroda M, et al: Interleukin-8 expression in the esophageal mucosa of patients with gastroesophageal reflux disease. Scand J Gastroenterol 2004;39:816–822.

8 Aihara M, Azuma A, Takizawa H, et al: Molecular analysis of suppression of interleukin-8 production by rebamipide in *Helicobacter pylori*-stimulated gastric cancer cell lines. Dig Dis Sci 1998;43: 174S–180S.

9 Crabtree JE, Lindley IJ: Mucosal interleukin-8 and *Helicobacter pylori*-associated gastroduodenal disease. Eur J Gastroenterol Hepatol 1994;6(suppl 1):S33–S38.

10 Kajikawa H, Yoshida N, Katada K, et al: *Helicobacter pylori* activates gastric epithelial cells to produce interleukin-8 via protease-activated receptor 2. Digestion 2008;76:248–255.

11 Kassai K, Yoshikawa T, Yoshida N, et al: *Helicobacter pylori* water extract induces interleukin-8 production by gastric epithelial cells. Dig Dis Sci 1999;44: 237–242.

12 Kuroda M, Yoshida N, Ichikawa H, et al: Lansoprazole, a proton pump inhibitor, reduces the severity of indomethacin-induced rat enteritis. Int J Mol Med 2006;17:89–93.

13 Okuda T, Yoshida N, Takagi T, et al: CV-11974, angiotensin II type I receptor antagonist, reduces the severity of indomethacin-induced rat enteritis. Dig Dis Sci 2008;53:657–663.

14 Naito Y, Katada K, Takagi T, et al: Rosuvastatin reduces rat intestinal ischemia-reperfusion injury associated with the preservation of endothelial nitric oxide synthase protein. World J Gastroenterol 2006;12:2024–2030.

15 Naito Y, Takagi T, Ichikawa H, et al: A novel potent inhibitor of inducible nitric oxide inhibitor, ONO-1714, reduces intestinal ischemia-reperfusion injury in rats. Nitric Oxide 2004;10:170–177.

16 Tsuboi H, Naito Y, Katada K, et al: Role of the thrombin/protease-activated receptor 1 pathway in intestinal ischemia-reperfusion injury in rats. Am J Physiol Gastrointest Liver Physiol 2007;292: G678–G683.

17 Anezaki K, Asakura H, Honma T, et al: Correlations between interleukin-8, and myeloperoxidase or luminol-dependent chemiluminescence in inflamed mucosa of ulcerative colitis. Intern Med 1998;37: 253–258.

18 Arai F, Takahashi T, Furukawa K, et al: Mucosal expression of interleukin-6 and interleukin-8 messenger RNA in ulcerative colitis and in Crohn's disease. Dig Dis Sci 1998;43:2071–2079.

19 Mazzucchelli L, Hauser C, Zgraggen K, et al: Expression of interleukin-8 gene in inflammatory bowel disease is related to the histological grade of active inflammation. Am J Pathol 1994;144:997–1007.

20 Ando T, Kusugami K, Ohsuga M, et al: Interleukin-8 activity correlates with histological severity in *Helicobacter pylori*-associated antral gastritis. Am J Gastroenterol 1996;91:1150–1156.

21 Watanabe T, Higuchi K, Kobata A, et al: Non-steroidal anti-inflammatory drug-induced small intestinal damage is Toll-like receptor 4 dependent. Gut 2008;57:181–187.

22 Fort MM, Mozaffarian A, Stover AG, et al: A synthetic TLR4 antagonist has anti-inflammatory effects in two murine models of inflammatory bowel disease. J Immunol 2005;174:6416–6423.

23 Ando A, Ogawa A, Bamba S, et al: Interaction between interleukin-17-producing CD4+ T cells and colonic subepithelial myofibroblasts: what are they doing in mucosal inflammation? J Gastroenterol 2007;42(Suppl 17):29–33.

24 Fujino S, Andoh A, Bamba S, et al: Increased expression of interleukin 17 in inflammatory bowel disease. Gut 2003;52:65–70.

25 Harrington LE, Mangan PR, Weaver CT: Expanding the effector CD4 T-cell repertoire: the Th17 lineage. Curr Opin Immunol 2006;18:349–356.

26 Mangan PR, Harrington LE, O'Quinn DB, et al: Transforming growth factor-beta induces development of the T(H)17 lineage. Nature 2006;441:231–234.

27 Park H, Li Z, Yang XO, et al: A distinct lineage of CD4 T cells regulates tissue inflammation by producing interleukin 17. Nat Immunol 2005;6:1133–1141.

28 Witowski J, Ksiazek K, Jorres A: Interleukin-17: a mediator of inflammatory responses. Cell Mol Life Sci 2004;61:567–579.

29 Kayo S, Ikura Y, Suekane T, et al: Close association between activated platelets and neutrophils in the active phase of ulcerative colitis in humans. Inflamm Bowel Dis 2006;12:727–735.

30 Danese S, de la Motte C, Sturm A, et al: Platelets trigger a CD40-dependent inflammatory response in the microvasculature of inflammatory bowel disease patients. Gastroenterology 2003;124:1249–1264.

31 Koutroubakis IE, Theodoropoulou A, Xidakis C, et al: Association between enhanced soluble CD40 ligand and prothrombotic state in inflammatory bowel disease. Eur J Gastroenterol Hepatol 2004;16:1147–1152.

32 Ludwiczek O, Kaser A, Tilg H: Plasma levels of soluble CD40 ligand are elevated in inflammatory bowel diseases. Int J Colorectal Dis 2003;18:142–147.

33 Danese S, Fiocchi C: Platelet activation and the CD40/CD40 ligand pathway: mechanisms and implications for human disease. Crit Rev Immunol 2005;25:103–121.

34 Vowinkel T, Anthoni C, Wood KC, et al: CD40-CD40 ligand mediates the recruitment of leukocytes and platelets in the inflamed murine colon. Gastroenterology 2007;132:955–965.

35 Zarbock A, Polanowska-Grabowska RK, Ley K: Platelet-neutrophil-interactions: linking hemostasis and inflammation. Blood Rev 2007;21:99–111.

36 Hofman V, Ricci V, Galmiche A, et al: Effect of *Helicobacter pylori* on polymorphonuclear leukocyte migration across polarized T84 epithelial cell monolayers: role of vacuolating toxin VacA and cag pathogenicity island. Infect Immun 2000;68:5225–5233.

37 Blake KM, Carrigan SO, Issekutz AC, et al: Neutrophils migrate across intestinal epithelium using beta2 integrin (CD11b/CD18)-independent mechanisms. Clin Exp Immunol 2004;136:262–268.

38 Carrigan SO, Weppler AL, Issekutz AC, et al: Neutrophil differentiated HL-60 cells model Mac-1 (CD11b/CD18)-independent neutrophil transepithelial migration. Immunology 2005;115:108–117.

39 Parkos CA, Colgan SP, Bacarra AE, et al: Intestinal epithelia (T84) possess basolateral ligands for CD11b/CD18-mediated neutrophil adherence. Am J Physiol 1995;268:C472–C479.

40 Weissmuller T, Campbell EL, Rosenberger P, et al: PMNs facilitate translocation of platelets across human and mouse epithelium and together alter fluid homeostasis via epithelial cell-expressed ecto-NTPDases. J Clin Invest 2008;118:3682–3692.

41 Yamamoto S, Shimizu S, Kiyonaka S, et al: TRPM2-mediated Ca^{2+} influx induces chemokine production in monocytes that aggravates inflammatory neutrophil infiltration. Nat Med 2008;14:738–747.

42 Yoshida N, Granger DN, Evans DJ Jr, et al: Mechanisms involved in *Helicobacter pylori*-induced inflammation. Gastroenterology 1993;105:1431–1440.

43 Montecucco C, de Bernard M: Molecular and cellular mechanisms of action of the vacuolating cytotoxin (VacA) and neutrophil-activating protein (HP-NAP) virulence factors of *Helicobacter pylori*. Microbes Infect 2003;5:715–721.

44 Evans DJ, Evans DG, Takemura T, et al: Characterization of a *Helicobacter pylori* neutrophil-activating protein. Infection Immunity 1995;63:2213–2220.

45 Naito Y, Yoshikawa T: Molecular and cellular mechanisms involved in *Helicobacter pylori*-induced inflammation and oxidative stress. Free Radic Biol Med 2002;33:323–336.

46 Satin B, Del Giudice G, Della Bianca V, et al: The neutrophil-activating protein (HP-NAP) of *Helicobacter pylori* is a protective antigen and a major virulence factor. J Exp Med 2000;191:1467–1476.

47 Polenghi A, Bossi F, Fischetti F, et al: The neutrophil-activating protein of *Helicobacter pylori* crosses endothelia to promote neutrophil adhesion in vivo. J Immunol 2007;178:1312–1320.

48 Aihara M, Tsuchimoto D, Takizawa H, et al: Mechanisms involved in *Helicobacter pylori*-induced interleukin-8 production by a gastric cancer cell line, MKN45. Infect Immun 1997;65:3218–3224.

49 Rieder G, Hatz RA, Moran AP, et al: Role of adherence in interleukin-8 induction in *Helicobacter pylori*-associated gastritis. Infect Immun 1997;65:3622–3630.

50 Sharma SA, Tummuru MK, Blaser MJ, et al: Activation of IL-8 gene expression by *Helicobacter pylori* is regulated by transcription factor nuclear factor-kappa B in gastric epithelial cells. J Immunol 1998;160:2401–2407.

51 Mai UA, Perez-Perez GI, Allen JB, et al: Surface proteins from *Helicobacter pylori* exhibit chemotactic activity for human leukocytes and are present in gastric mucosa. J Exp Med 1992;175:517–525.

52 Santucci L, Fiorucci S, Giansanti M, et al: Pentoxifylline prevents indomethacin induced acute gastric mucosal damage in rats: role of tumour necrosis factor alpha. Gut 1994;35:909–915.
53 Santucci L, Fiorucci S, Di MF, et al: Role of tumor necrosis factor alpha release and leukocyte margination in indomethacin-induced gastric injury in rats. Gastroenterology 1995;108:393–401.
54 Fiorucci S, Antonelli E, Migliorati G, et al: TNFalpha processing enzyme inhibitors prevent aspirin-induced TNFalpha release and protect against gastric mucosal injury in rats. Aliment Pharmacol Ther 1998;12:1139–1153.
55 Yoshikawa T, Takano H, Naito Y, et al: Augmentative effects of tumor necrosis factor-alpha (human, natural type) on polymorphonuclear leukocyte-derived superoxide generation induced by various stimulants. Int J Immunopharmacol 1992;14:1391–1398.
56 Kokura S, Wolf RE, Yoshikawa T, et al: T-lymphocyte-derived tumor necrosis factor exacerbates anoxia-reoxygenation-induced neutrophil-endothelial cell adhesion. Circ Res 2000;86:205–213.
57 Asako H, Kubes P, Wallace J, et al: Indomethacin-induced leukocyte adhesion in mesenteric venules: role of lipoxygenase products. Am J Physiol 1992; 262:G903-G908.
58 Miura S, Suematsu M, Tanaka S, et al: Microcirculatory disturbance in indomethacin-induced intestinal ulcer. Am J Physiol 1991;261: G213–G219.
59 Hanai H, Takeuchi K, Iida T, et al: Relationship between fecal calprotectin, intestinal inflammation, and peripheral blood neutrophils in patients with active ulcerative colitis. Dig Dis Sci 2004;49:1438–1443.
60 McCarthy DA, Rampton DS, Liu YC: Peripheral blood neutrophils in inflammatory bowel disease: morphological evidence of in vivo activation in active disease. Clin Exp Immunol 1991;86:489–493.
61 D'Odorico A, D'Inca R, Mestriner C, et al: Influence of disease site and activity on peripheral neutrophil function in inflammatory bowel disease. Dig Dis Sci 2000;45:1594–1600.
62 Isaacs KL, Sartor RB, Haskill S: Cytokine messenger RNA profiles in inflammatory bowel disease mucosa detected by polymerase chain reaction amplification. Gastroenterology 1992;103:1587–1595.
63 Carlson M, Raab Y, Seveus L, et al: Human neutrophil lipocalin is a unique marker of neutrophil inflammation in ulcerative colitis and proctitis. Gut 2002;50:501–506.
64 Kanai T, Hibi T, Watanabe M: The logics of leukocytapheresis as a natural biological therapy for inflammatory bowel disease. Expert Opin Biol Ther 2006;6:453–466.
65 Shimoyama T, Sawada K, Hiwatashi N, et al: Safety and efficacy of granulocyte and monocyte adsorption apheresis in patients with active ulcerative colitis: a multicenter study. J Clin Apher 2001;16:1–9.
66 Hanai H, Watanabe F, Takeuchi K, et al: Leukocyte adsorptive apheresis for the treatment of active ulcerative colitis: a prospective, uncontrolled, pilot study. Clin Gastroenterol Hepatol 2003;1:28–35.
67 Suzuki Y, Yoshimura N, Saniabadi AR, et al: Selective granulocyte and monocyte adsorptive apheresis as a first-line treatment for steroid naive patients with active ulcerative colitis: a prospective uncontrolled study. Dig Dis Sci 2004;49:565–571.
68 Fukuda Y, Matsui T, Suzuki Y, et al: Adsorptive granulocyte and monocyte apheresis for refractory Crohn's disease: an open multicenter prospective study. J Gastroenterol 2004;39:1158–1164.
69 Ljung T, Thomsen OO, Vatn M, et al: Granulocyte, monocyte/macrophage apheresis for inflammatory bowel disease: the first 100 patients treated in Scandinavia. Scand J Gastroenterol 2007;42:221–227.
70 Naito Y, Takano H, Yoshikawa T: Oxidative stress-related molecules as a therapeutic target for inflammatory and allergic diseases. Curr Drug Targets Inflamm Allergy 2005;4:511–515.
71 Yoshikawa T, Naito Y: What is oxidative stress? JMAJ 2002;45:271–276.
72 Naito Y, Takagi T, Handa O, et al: Role of superoxide and lipid peroxidation in the pathogenesis of dextran sulfate sodium-colitis in mice. ITE Lett 2001a; 2:663–667.
73 Yoshikawa T, Naito Y, Kishi A, et al: Role of active oxygen, lipid peroxidation, and antioxidants in the pathogenesis of gastric mucosal injury induced by indomethacin in rats. Gut 1993;34:732–737.
74 Yoshikawa T, Ueda S, Naito Y, et al: Role of oxygen-derived free radicals in gastric mucosal injury induced by ischemia or ischemia-reperfusion in rats. Free Radic Res Commun 1989;7:285–291.
75 Naito Y, Takagi T, Matsuyama K, et al: Effect of a novel water-soluble vitamin E derivative, 2-(a-D-glucopyranoyl)methyl-2,5,7,8-tetramethylchroman-6-ol, on dextran sulfate sodium-inducd colitis in mice. J Clin Biochem Nutr 2002;31:59–67.
76 Uchida K: Protein-bound 4-hydroxy-2-nonenal as a marker of oxidative stress. J Clin Biochem Nutr 2005;36:1–10.
77 Dwivedi S, Sharma A, Patrick B, et al: Role of 4-hydroxynonenal and its metabolites in signaling. Redox Rep 2007;12:4–10.
78 Leonarduzzi G, Robbesyn F, Poli G: Signaling kinases modulated by 4-hydroxynonenal. Free Radic Biol Med 2004;37:1694–1702.

79 Bartsch H, Nair J: Accumulation of lipid peroxidation-derived DNA lesions: potential lead markers for chemoprevention of inflammation-driven malignancies. Mutat Res 2005;591:34–44.

80 Nair J, Gansauge F, Beger H, et al: Increased etheno-DNA adducts in affected tissues of patients suffering from Crohn's disease, ulcerative colitis, and chronic pancreatitis. Antioxid Redox Signal 2006;8: 1003–1010.

81 Kimura H, Mukaida M, Kuwabara K, et al: 4-Hydroxynonenal modifies IgA in rat intestine after lipopolysaccharide injection. Free Radic Biol Med 2006;41:973–978.

82 Biasi F, Vizio B, Mascia C, et al: c-Jun N-terminal kinase upregulation as a key event in the proapoptotic interaction between transforming growth factor-beta1 and 4-hydroxynonenal in colon mucosa. Free Radic Biol Med 2006;41:443–454.

83 Grisham MB, Jefferson MM, Thomas EL: Role of monochloramine in the oxidation of erythrocyte hemoglobin by stimulated neutrophils. J Biol Chem 1984;259:6757–6765.

84 Tamai H, Kachur JF, Baron DA, et al: Monochloramine, a neutrophil-derived oxidant, stimulates rat colonic secretion. J Pharmacol Exp Ther 1991;257:887–894.

85 Thomas EL, Grisham MB, Jefferson MM: Preparation and characterization of chloramines. In: DiSabato G., Everse J. (Eds.), Methods in Enzymology, Vol132. Academic Press, Orlando, 1986, pp 569–584.

86 Suzuki M, Miura S, Suematsu M, et al: *Helicobacter pylori*-associated ammonia production enhances neutrophil-dependent gastric mucosal cell injury. Am J Physiol 1992;263:G719–G725.

87 Naito Y, Yoshikawa T, Fujii T, et al: Monochloramine-induced cell growth inhibition and apoptosis in a rat gastric mucosal cell line. J Clin Gastroenterol 1997;25:S179–S185.

88 Suzuki H, Mori M, Suzuki M, et al: Extensive DNA damage induced by monochloramine in gastric cells. Cancer Lett 1997;115:243–248.

89 Winterbourn CC, Kettle AJ: Biomarkers of myeloperoxidase-derived hypochlorous acid. Free Radic Biol Med 2000;29:403–409.

90 Gujral JS, Hinson JA, Jaeschke H: Chlorotyrosine protein adducts are reliable biomarkers of neutrophil-induced cytotoxicity in vivo. Comp Hepatol 2004;3(Suppl 1):S48.

91 Ryu JK, Tran KC, McLarnon, JG: Depletion of neutrophils reduces neuronal degeneration and inflammatory responses induced by quinolinic acid in vivo. Glia 2007;55:439–451.

92 Shao B, Oda MN, Bergt C, et al: Myeloperoxidase impairs ABCA1-dependent cholesterol efflux through methionine oxidation and site-specific tyrosine chlorination of apolipoprotein A-I. J Biol Chem 2006;281:9001–9004.

93 Kettle AJ, Chan T, Osberg I, et al: Myeloperoxidase and protein oxidation in the airways of young children with cystic fibrosis. Am J Respir Crit Care Med 2004;170:1317–1323.

94 Jiang Q, Blount BC, Ames BN: 5-Chlorouracil, a marker of DNA damage from hypochlorous acid during inflammation. A gas chromatography-mass spectrometry assay. J Biol Chem 2003;278:32834–32840.

95 Asahi T, Kondo H, Masuda M, et al: Chemical and immunochemical detection of 8-halogenated deoxyguanosines at early stage inflammation. J Biol Chem 2010;285:9282–9291.

96 Kawai Y, Morinaga H, Kondo H, et al: Endogenous formation of novel halogenated 2′-deoxycytidine. Hypohalous acid-mediated DNA modification at the site of inflammation. J Biol Chem 2004;279: 51241–51249.

97 Naito Y, Takagi T, Ishikawa T, et al: The inducible nitric oxide synthase inhibitor ONO-1714 blunts dextran sulfate sodium colitis in mice. Eur J Pharmacol 2001b;412:91–99.

98 Mazzon E, Muia C, Paola RD, et al: Green tea polyphenol extract attenuates colon injury induced by experimental colitis. Free Radic Res 2005;39:1017–1025.

99 Cuzzocrea S, Mazzon E, Dugo L, et al: Protective effects of M40403, a superoxide dismutase mimetic, in a rodent model of colitis. Eur J Pharmacol 2001; 432:79 89.

100 Kruidenier L, Kuiper I, LamersCB, et al: Intestinal oxidative damage in inflammatory bowel disease: semi-quantification, localization, and association with mucosal antioxidants. J Pathol 2003;201:28–36.

101 Naito Y, Takagi T, Okada H, et al: Expression of inducible nitric oxide synthase and nitric oxide-modified proteins in *Helicobacter pylori*-associated atrophic gastric mucosa. J Gastroenterol Hepatol 2008;23(suppl 2):S250–S257.

102 Singer I, Kawka DW, Scott S, et al: Expression of inducible nitric oxide synthase and nitrotyrosine in colonic epithelium in inflammatory bowel disease. Gastroenterology 1996;111:871–885.

103 Keshavarzian A, Banan A, Farhadi A, et al: Increases in free radicals and cytoskeletal protein oxidation and nitration in the colon of patients with inflammatory bowel disease. Gut 2003;52:720–728.

104 Schmid M, Fellermann K, Fritz P, et al: Attenuated induction of epithelial and leukocyte serine antiproteases elafin and secretory leukocyte protease inhibitor in Crohn's disease. J Leukoc Biol 2007;81: 907–915.

105 Morohoshi Y, Matsuoka K, Chinen H, et al: Inhibition of neutrophil elastase prevents the development of murine dextran sulfate sodium-induced colitis. J Gastroenterol 2006;41:318–324.

106 Young RE, Thompson RD, Larbi KY, et al: Neutrophil elastase (NE)-deficient mice demonstrate a nonredundant role for NE in neutrophil migration, generation of proinflammatory mediators, and phagocytosis in response to zymosan particles in vivo. J Immunol 2004;172:4493–4502.

Yuji Naito
Molecular Gastroenterology and Hepatology
Kyoto Prefectural University of Medicine
Kamigyo-ku, Kyoto 602-8566 (Japan)
Tel. +81 75 251 5519, Fax +81 75 251 0710, E-Mail ynaito@koto.kpu-m.ac.jp

Naito Y, Suematsu M, Yoshikawa T (eds): Free Radical Biology in Digestive Diseases.
Front Gastrointest Res. Basel, Karger, 2011, vol 29, pp 55–63

Oxidative DNA Damage and Carcinogenesis

Li Jiang · Shinya Toyokuni

Department of Pathology and Biological Responses, Nagoya University Graduate School of Medicine, Nagoya, Japan

Abstract

Carcinogenesis follows multi-step processes involving both genetic alteration and increased cell proliferation. Oxidative stress can occur via overproduction of reactive oxygen and nitrogen species through either endogenous or exogenous insults. Oxidative stress is always associated with inflammation, radiation, reperfusion, and iron overload. Epidemiological observations have shown that oxidative stress is one of the major pathologic mechanisms for cancer, the top-ranked cause of human mortality worldwide. During carcinogenesis, the unregulated or prolonged production of cellular oxidants has been linked to mutation through generation of oxidative DNA damage. Furthermore, signal transduction pathways are activated by reactive species, and they lead to the transcription of genes involved in cellular growth regulatory and stress protection pathways. This review examines the involvement of oxidative stress in the carcinogenic process starting from its history, and presents future perspectives.

Oxidative stress is associated with a plethora of pathological phenomena, including infection, inflammation, ultraviolet and γ-irradiation, overload of transition metals and exposure to certain chemical agents as well as ischemia-reperfusion injury [1]. Free radicals and other reactive species are constantly generated in vivo and cause oxidative damage to DNA at a rate that is probably a significant contributor to carcinogenesis. All living cells are always exposed to a limited level of reactive oxygen species (ROS). However, if ROS levels increase, oxidative stress occurs, which results in the formation of oxygen-derived oxidants, and in turn increases the rates of molecular damage in cells. Several forms of genomic DNA damage are caused by ROS, including chromatin cross-linking, chromosome deletion, DNA strand breaks and base oxidation. Agents that decrease oxidative DNA damage should thus decrease the risk of cancer development. That is, oxidative DNA damage can be a 'biomarker' for identifying persons at risk (for dietary or genetic reasons, or both) of developing cancer and for suggesting how the diets of these persons could be modified to decrease that risk [2].

The History of Free Radical Chemistry

The history of free radical chemistry dates back to 116 years ago. Fenton, a British chemist, reported in 1894 that ferrous sulfate and hydrogen peroxide cause the oxidation of tartaric acid, resulting in a beautiful violet color on the addition of caustic alkali [3]. This was the basis for the discovery of the Fenton reaction, which produces hydroxyl radicals, presently known as the most reactive endogenous chemical species in living cells. At present, it is well established that iron overload is associated with carcinogenesis [4]. In 1954, Harman proposed a hypothesis of 'free radicals as a cause of aging' for the first time [5]. A decade was necessary to integrate this idea into science, when McCord and Fridovich discovered superoxide dismutase, the first enzyme whose substrate is a free radical, in 1969 [6]. For the next two decades, studies based on the hypothesis that reactive species are damaging biomolecules were planned, and those efforts led to pioneering works on antioxidants, antioxidant enzymes, and protein oxidation, as well as DNA modification. Thus, the concept of 'oxidative stress' in the original sense was established [7].

During the 1990s, a number of scientists newly joined this research area using molecular biology techniques, and low levels of reactive oxygen and nitrogen species were established as signaling molecules in the cell [8, 9]. Multiple repair enzymes for oxidative DNA damage were also identified in this period [10].

Reactive Oxygen Species

ROS are generated in mitochondria of normal mammalian cells as byproducts of routine respiration process as well as in other subcellular locations as a function of biochemical reactions using oxygen and peroxides. ROS at high levels are toxic to the cell, but at low levels, ROS have physiological functions, including activation and modulation of signal transduction pathways, modulation of activities of redox-sensitive transcription factors, and regulation of mitochondrial enzyme activities. Levels of ROS are reduced by antioxidant defenses, but increased with transition metals such as iron or copper and by exogenous agents such as ionizing radiation or ozone [11].

Antioxidant Systems

To protect against toxic effects of ROS and to modulate physiological effects of ROS, the cell has developed an intricately regulated antioxidant defense system. The antioxidant enzyme system is complex but designed to compensate each other, being composed of small-molecular-weight antioxidant compounds (vitamins E, C and A; plant polyphenols from foods, etc.); primary (superoxide dismutases, catalase, glutathione peroxidase) and secondary antioxidant enzymes (enzymes such as glutathione

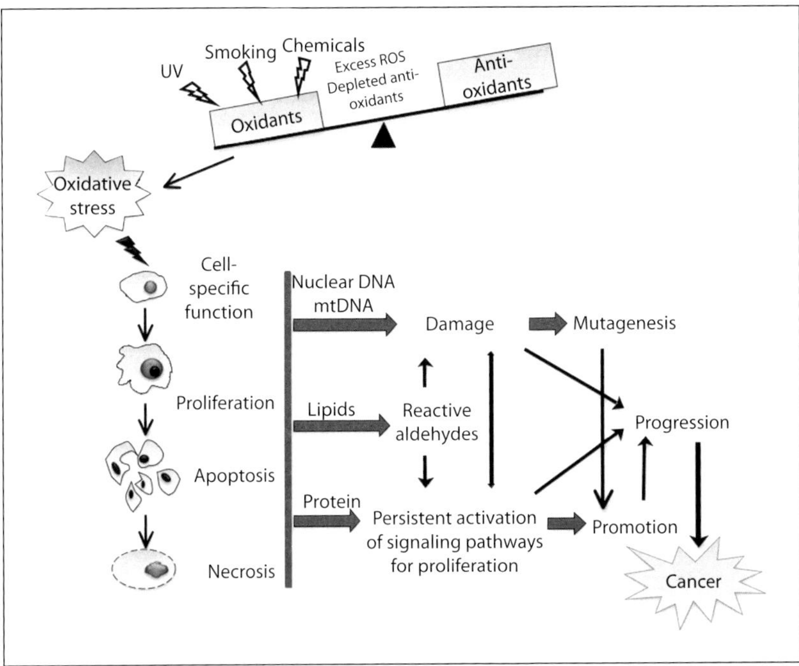

Fig. 1. Biological significance of oxidative stress in carcinogenesis.

reductase and glucose-6-phosphate dehydrogenase), and the glutathione, glutaredoxin, and thioredoxin systems. Protein and DNA repair enzymes are considered part of the antioxidant system. Proteins that sequester metals such as ferritin are important in modulating cellular redox state. Nitric oxide modulates levels of ROS in part by its reaction with superoxide anion. Finally, proteins involved in response to cell stressors, such as the heat shock protein system [12], are important in the prevention of oxidative damage. Each component of the antioxidant system is specifically compartmentalized to specific subcellular locations.

In summary, oxidative stress depends on a balance between the load of reactive species and the antioxidant systems responsible for them. The latter contains (a) protective mechanisms for the initiation of the free radical reaction; (b) what are called 'anti-oxidants' that efficiently retard the free radical reaction after its initiation, and lastly (c) repair mechanisms for modified or damaged biomolecules.

Oxidative Stress and DNA

DNA is susceptible to many kinds of damage and one of these is oxidative damage. This type of damage can build up over time and overwhelm repair systems, leading to health problems and eventually, disease. ROS are a consequence of normal body processes such as metabolism. They are also a consequence of interactions with toxic sources such as cancer-promoting substances, certain drugs and radiation. DNA oxidative damage from ROS is a common type of damage faced by cells, and it can lead to many different kinds of mutations in DNA. These range from lesions on DNA and breaks in DNA strands to faulty links and base gaps in sequences. DNA oxidative damage and its effects can be extremely detrimental to the cell. Our understanding of precisely how these effects occur is important to prevent the damage or to look at how to ease the rate at which oxidative reactions cause damage to the body. Another valid focus is to find ways to measure the level of damage before it overwhelms the body and leads to disease. At present, many different types of DNA lesions have been identified in the cells in vivo under different conditions [13]. This information has allowed researchers to understand just how DNA oxidative damage is implicated in a number of diseases including arthritis, heart disease, ageing, gastrointestinal disorders, multiple sclerosis, degenerative neurological diseases and cancer [14].

Oxidative DNA Damage and Carcinogenesis

ROS can directly induce many forms of damage in DNA, including single- or double-stranded DNA breaks, purine, pyrimidine, or deoxyribose modifications, and DNA-protein cross-links, and deoxyribose oxidation [15, 16]. Persistent DNA damage can result in either arrest or induction of transcription, induction of signal transduction pathways, replication errors, activation of cell signaling pathways, aberrant cell proliferation and genomic instability, all of which are seen during carcinogenesis.

Recently, asbestos exposure has received much social attention when it was discovered to have a strong association with the development of a rare tumor, known as malignant mesothelioma [17]. Asbestos is a variety of fibrous natural stone with its common major content as silicate oxide, and has been used from the Egyptian era because of cheapness and its heat and friction durability. In Japan, Gen-nai Hiraga produced a heat-resistant textile in 1764 by the use of asbestos found in the mountainous area of Chichibu in the suburbs of Edo (the ancient name of Tokyo). The toxicity of asbestos fiber by inhalation is characterized by a number of processes, among which ROS production is thought to be the most important one. Highly reactive ROS such as the hydroxyl radical can be produced through Fenton-type reaction catalyzed by iron component or impurities present on the surface [18]. ROS are also produced in the lungs by the chronic inflammatory reaction produced by the prolonged phagocytic activity of macrophages against the bio-persistent fibers.

The hydroxyl radical in particular has been shown to produce a number of oxidized DNA lesions. The reactivity of the hydroxyl radical is such that its migration in the cell is limited and thus reacts quickly with cellular components. For the hydroxyl radical to react and oxidize DNA, it must be generated adjacent to the nucleic acid materials. H_2O_2, a precursor to hydroxyl radical, is less reactive and more readily diffusible and thus more likely to be involved in the formation of oxidized bases. Peroxynitrite, another strong cellular oxidant, is formed from the coupling of nitric oxide and superoxide [19]. As with H_2O_2, nitric oxide is diffusible between cells and is taken up into cells. Equally important to the induction of mutation by ROS is the fact that nitric oxide and superoxide are produced in activated macrophages, and as such, it is likely that peroxynitrite is formed in proximity to these cells. The DNA damaging capability of peroxynitrite [20] may therefore help to explain the reported association between inflammation and mutation. Singlet oxygen associated with ultraviolet radiation/visible light as well as hypochlorous acid is another source of oxidative DNA damage [21].

Oxidation of guanine at the C8 position results in the formation of 8-hydroxy-2′-deoxyguanosine (8-OHdG), probably the most studied oxidative DNA adduct [22]. This oxidative DNA lesion results in site-specific mutagenesis, is mutagenic in bacterial and mammalian cells, and produces G to T transversions [23] that are widely found in mutated oncogenes and tumor suppressor genes [24]. In addition, hydroxyl radicals can react with dGTP in the nucleotide pool to form 8-OH-dGTP. Therefore, it is postulated that during DNA replication, 8-OH-dGTP in the nucleotide pool will be incorporated into DNA opposite dA on the template strand, resulting in A to C transversions [25]. 8-OHdG also produces dose-related increases in cellular transformation, which can be prevented by antioxidants, further supporting the role of 8-OHdG in the carcinogenic process [10]. Other oxidative DNA lesions, such as 8-oxo-adenine, thymine glycol, 5-hydroxy-deoxycytidine, as well as several uracil analogs, have been shown to be mutagenic.

Oxidized DNA bases appear to be mutagenic and capable of inducing mutations that are commonly observed in human neoplasia. At high cytotoxic concentrations, asbestos fibers caused formation of the premutagenic lesion, 8-OHdG [18], which might be rapidly repaired and/or released into cell medium by the injured mesothelial cells. While genomic alterations are an essential feature in the causation of cancer, large chromosomal deletions by asbestos are associated with its clastogenicity and largely associated with cell death [26].

Cancer development is a multi-factorial, multi-step process characterized by the cumulative action of additive events occurring in a single cell [27]. It is generally described by three stages: initiation, promotion and progression. Mutations, oncogene activation, genomic instability, and epigenetic factors dysregulate gene expression profiles and disrupt the molecular networks, providing tumor cells with sufficient diversity and proliferative advantage to evolve in the environment of original tissue and eventually to metastasize. Oxidative DNA damage can occur both in the nucleus and mitochondria. ROS are potential carcinogens because they can facilitate all three stages of tumor formation. As initiation, a non-lethal mutation in DNA is followed by at least one round of DNA replication to fix the genomic alteration and produce an initiated cell. The promotion stage requires the continuous presence of the tumor promotion stimulus, and therefore it is a reversible process. This stage may eventually result in the formation of an identifiable focal lesion that is characterized by the clonal expansion of initiated cells via the induction of cell proliferation and/or inhibition of apoptosis. Production of ROS during this stage of carcinogenesis is the main line of ROS-related protumor effect as many carcinogens have a strong inhibiting effect on cellular antioxidant defense systems such as superoxide dismutase, catalase, and glutathione. Low sustained level of oxidative stress in this state may also stimulate cell division resulting thus in tumor growth. The final stage of the carcinogenic process is progression. This stage involves cellular and molecular changes that occur from the preneoplastic to the neoplastic state and is irreversible as it includes accumulation of additional genetic damage, genetic instability, and angiogenesis.

Oxygenomics

We have been studying the molecular pathways of oxidative stress-induced carcinogenesis using ferric nitrilotriacetate-induced renal cell carcinoma [28] and iron saccharate-induced mesothelioma in rats [29]. We found many target genes in these carcinogenesis models. Of note was the fact that *CDKN2A/2B* ($p16^{INK4A}$ and $p15^{INK4B}$ tumor suppressor genes) are the most common target genes, with most of them deleted homozygously [30, 31]. Based on these observations, we believe that there are fragile sites against oxidative stress that probably depend on the state of the chromatin, whether open or closed, the site of ROS generation as well as chromosome territory. Studies are in progress to find a common principal by the use of DNA immunoprecipitation strategy [32], and we have proposed this research area as 'oxygenomics' [33].

Conclusions

It is reasonable to speculate that given the multiple pathways for its repair, oxidative DNA damage is likely to play an important role in carcinogenesis. Indeed, it seems that ROS and oxidative DNA damage are 'omnipresent' in cancer; for researchers, this means there is no limit to the conditions in which oxidative stress may be studied. However, the mere presence of damage is not a proof of causative link, although given the close link between ROS formation and oxidative DNA damage and the importance of DNA damage and mutation in carcinogenesis, it is not a large leap of intuition to link oxidative DNA lesions and cancer.

A linkage between an increase in cellular ROS and the pathogenesis of several chronic diseases including cancer has been established. Cellular oxidants can be generated from endogenous as well as exogenous sources. When the antioxidant control mechanisms are exhausted or overrun, the cellular redox potential shifts toward an oxidative stress, in turn, increasing the potential for damage to cellular nucleic acids, lipid, and protein. Unrepaired damage to DNA may result in mutation, provided cell replication ensues prior to repair of modified bases. Although importance has been established for the role of oxidative nuclear DNA damage in neoplasia, formation of mitochondrial DNA damage, mutation and alteration of the mitochondrial genomic function also appear to contribute to the process of carcinogenesis.

The role of ROS in the regulation of cell growth is complex, being cell specific and dependent upon the form of oxidants as well as the concentration of the particular ROS. The modification of gene expression by ROS has direct effects on cell proliferation and apoptosis through the activation of transcription factors including mitogen-activated protein kinase, AP-1, and NF-κB pathways [8]. Oxidant-mediated AP-1 activation results in enhanced expression of cyclin D1 and cyclin-dependent kinase, which in turn promotes entry into mitosis and cell division. Likewise, ROS function as second messengers involved in activation of NF-κB by tumor necrosis factor-α and other cytokines. Recently, Nrf2-Keap1 antioxidant system attracted much attention since cancer cells hijack this system [34]. Gene expression is also controlled by the methylation status of genomic DNA. Recent evidence shows that ROS as well as oxidative DNA damage modulate DNA methylation patterns, and may contribute to the multistage carcinogenesis process. DNA damage, mutation, and altered gene expression are all required participants in the process of carcinogenesis. Although these events may be derived by different mechanisms, the common observation is the involvement of cellular oxidants in the neoplastic development.

Acknowledgements

This work was supported in part by a MEXT grant (Special Coordination Funds for Promoting Science and Technology), a grant of Long-range Research Initiative by Japan Chemical Industry Association, a Grant from Takeda Science Foundation, a Grant-in-Aid for Cancer Research from the Ministry of Health, Labour and Welfare of Japan, and a Grant-in-Aid from the Ministry of Education, Culture, Sports, Science and Technology of Japan.

References

1 Toyokuni S: Reactive oxygen species-induced molecular damage and its applicaton in pathology. Pathol Int 1999;49:91–102.

2 Arai S, Osawa T, Ohigashi H, Yoshikawa M, Kaminogawa S, Watanabe M, Ogawa T, Okubo K, Watanabe S, Nishino H, Shinohara K, Esashi T, Hirahara T: A mainstay of functional food science in Japan – history, present status, and future outlook. Biosci Biotechnol Biochem 2001;65:1–13.

3 Fenton HJH: Oxidation of tartaric acid in presence of iron. J Chem Soc 1894;65:899–910.

4 Toyokuni S: Role of iron in carcinogenesis: cancer as a ferrotoxic disease. Cancer Sci 2009;100:9–16.

5 Harman D: Aging: a theory based on free radical and radiation chemistry. J Gerontol 1956;11:298–300.

6 McCord JM, Fridovich I: Superoxide dismutase: an enzymatic function for erythrocuprein (hemocuprein). J Biol Chem 1969;244:6049–6055.

7 Toyokuni S: Novel aspects in free radical chemistry and biology. Arch Biochem Biophys 2003;417: 2–2.

8 Sen C, Packer L: Antioxidant and redox regulation of gene transcription. FASEB J 1996;10:709–720.

9 Itoh K, Chiba T, Takahashi S, Ishii T, Igarashi K, Katoh Y, Oyake T, Hayashi N, Satoh K, Hatayama I, Yamamoto M, Nabeshima Y: An Nrf2/small Maf heterodimer mediates the induction of phase II detoxifying enzyme genes through antioxidant response elements. Biochem Biophys Res Commun 1997;236:313–322.

10 Nakabeppu Y, Tsuchimoto D, Furuichi M, Sakumi K: The defense mechanisms in mammalian cells against oxidative damage in nucleic acids and their involvement in the suppression of mutagenesis and cell death. Free Radic Res 2004;38:423–429.

11 Toyokuni S, Akatsuka S: Pathological investigation of oxidative stress in the post-genomic era. Pathol Int 2007;57:461–473.

12 Yamamoto H, Yamamoto Y, Yamagami K, Kume M, Kimoto S, Toyokuni S, Uchida K, Fukumoto M, Yamaoka Y: Heat-shock preconditioning reduces oxidative protein denaturation and ameliorates liver injury by carbon tetrachloride in rats. Res Exp Med (Berl) 2000;199:309–318.

13 Dizdaroglu M: Chemical determination of free radical-induced damage to DNA. Free Radic Biol Med 1991;10:225–242.

14 Evans MD, Dizdaroglu M, Cooke MS: Oxidative DNA damage and disease: induction, repair and significance. Mutat Res 2004;567:1–61.

15 Toyokuni S, Mori T, Dizdaroglu M: DNA base modifications in renal chromatin of Wistar rats treated with a renal carcinogen, ferric nitrilotriacetate. Int J Cancer 1994;57:123–128.

16 Toyokuni S, Mori T, Hiai H, Dizdaroglu M: Treatment of Wistar rats with a renal carcinogen, ferric nitrilotriacetate, causes DNA-protein cross-linking between thymine and tyrosine in their renal chromatin. Int J Cancer 1995;62:309–313.

17 Toyokuni S: Mechanisms of asbestos-induced carcinogenesis. Nagoya J Med Sci 2009;71:1–10.

18 Jiang L, Nagai H, Ohara H, Hara S, Tachibana M, Hirano S, Shinohara Y, Kohyama N, Akatsuka S, Toyokuni S: Characteristics and modifying factors of asbestos-induced oxidative DNA damage. Cancer Sci 2008;99:2142–2151.

19 Beckman J, Koppenol W: Nitric oxide, superoxide, and peroxynitrite: the good, the bad, and ugly. Am J Physiol 1996;271:C1424–C1437.

20 Szabo C, Ohshima H: DNA damage induced by peroxynitrite: subsequent biological effects. Nitric Oxide 1997;1:373–385.

21 Sies H, Menck CF: Singlet oxygen induced DNA damage. Mutat Res 1992;275:367–375.

22 Kasai H, Nishimura S: Hydroxylation of deoxyguanosine at the C-8 position by ascorbic acid and other reducing agents. Nucleic Acids Res 1984;12:2137–2145.

23 Shibutani S, Takeshita M, Grollman AP: Insertion of specific bases during DNA synthesis past the oxidation-damaged base 8-oxodG. Nature 1991;349:431–434.

24 Hussain SP, Harris CC: Molecular epidemiology of human cancer: contribution of mutation spectra studies of tumor suppressor genes. Cancer Res 1998;58:4023–4037.

25 Furuichi M, Yoshida MC, Oda H, Tajiri T, Nakabeppu Y, Tsuzuki T, Sekiguchi M: Genomic structure and chromosome location of the human mutT homologue MTH1 encoding 8-oxo-dGTPase for prevention of A:T to C:G transversion. Genomics 1994;24:485–490.

26 Hei TK, Piao CQ, He ZY, Vannais D, Waldren CA: Chrysotile fiber is a strong mutagen in mammalian cells. Cancer Res 1992;52:6305–6309.

27 Toyokuni S: Novel aspects of oxidative stress-associated carcinogenesis. Antioxid Redox Signal 2006;8:1373–1377.

28 Liu Y-T, Shang D-G, Akatsuka S, Ohara H, Dutta KK, Mizushima K, Naito Y, Yoshikawa T, Izumiya M, Abe K, Nakagama H, Noguchi N, Toyokuni S: Chronic oxidative stress causes amplification and overexpresson of ptprz1 protein tyrosine phosphatase to activate b-catenin pathway. Am J Pathol 2007;171:1978–1988.

29 Okada S, Hamazaki S, Toyokuni S, Midorikawa O: Induction of mesothelioma by intraperitoneal injections of ferric saccharate in male Wistar rats. Br J Cancer 1989;60:708–711.

30 Tanaka T, Iwasa Y, Kondo S, Hiai H, Toyokuni S: High incidence of allelic loss on chromosome 5 and inactivation of $p15^{INK4B}$ and $p16^{INK4A}$ tumor suppressor genes in oxystress-induced renal cell carcinoma of rats. Oncogene 1999;18:3793–3797.

31 Hu Q, Akatsuka S, Yamashita Y, Ohara H, Nagai H, Okazaki Y, Takahashi T, Toyokuni S: Homozygous deletion of *CDKN2A/2B* is a hallmark of iron-induced high-grade rat mesothelioma. Lab Invest 2010;90:360–373.

32 Akatsuka S, Aung TT, Dutta KK, Jiang L, Lee W-H, Liu Y-T, Onuki J, Shirase T, Yamasaki K, Ochi H, Naito Y, Yoshikawa T, Kasai H, Tominaga Y, Sakumi K, Nakabeppu Y, Kawai Y, Uchida K, Yamasaki A, Tsuruyama T, Yamada Y, Toyokuni S: Contrasting genome-wide distribution of 8-hydroxyguanine and acrolein-modified adenine during oxidative stress-induced renal carcinogenesis. Am J Pathol 2006;169:1328–1342.
33 Toyokuni S: Molecular mechanisms of oxidative stress-induced carcinogenesis: from epidemiology to oxygenomics. IUBMB Life 2008;60:441–447.
34 Ohta T, Iijima K, Miyamoto M, Nakahara I, Tanaka H, Ohtsuji M, Suzuki T, Kobayashi A, Yokota J, Sakiyama T, Shibata T, Yamamoto M, Hirohashi S: Loss of Keap1 function activates Nrf2 and provides advantages for lung cancer cell growth. Cancer Res 2008;68:1303–1309.

Prof. Shinya Toyokuni
Department of Pathology and Biological Responses
Nagoya University Graduate School of Medicine
65 Tsurumai-cho, Showa-ku
Nagoya, Aichi 466-8550 (Japan)
Tel. +81 52 744 2086, Fax +81 52 744 2091, E-Mail toyokuni@med.nagoya-u.ac.jp

Naito Y, Suematsu M, Yoshikawa T (eds): Free Radical Biology in Digestive Diseases.
Front Gastrointest Res. Basel, Karger, 2011, vol 29, pp 64–70

Plasma Marker of Oxidative Stress in Circulation and in Tissue

Yorihiro Yamamoto

School of Bioscience and Biotechnology, Tokyo University of Technology, Tokyo, Japan

Abstract

Since the reduced form of coenzyme Q10 ($CoQ10H_2$) is oxidized to its oxidized form (CoQ10) at early stage, the percentage of CoQ10 in total coenzyme Q10 (= CoQ10 + $CoQ10H_2$) is a sensitive plasma marker of oxidative stress in circulation. An increased oxidative stress was demonstrated in patients with hepatitis, cirrhosis, and hepatoma, in LEC rats (animal model of liver cancer), in newborn babies, in patients with Parkinson's disease, and in patients treated with direct percutaneous transluminal coronary angioplasty. Tissue oxidative damage results in increased hydrolyzed free fatty acids in bloodstream. One can assume a decreased content of polyunsaturated fatty acids (PUFA) and an increased content of oleic acid (18:1) and palmitoleic acid (16:1) since these monoenoic acids are produced by the action of stearoyl-CoA desaturase to compensate for the loss of PUFA. Such changes in the plasma were a useful marker of oxidative stress in tissue and were confirmed in rats with carbon tetrachloride poisoning, in LEC rats, in rats that received 2-hour middle cerebral artery occlusion-reperfusion, and in patients with hepatitis, cirrhosis, and hepatoma.

Oxidative stress is defined as a disturbance in the pro-oxidant-antioxidant balance in favor of the former. Inflammation, ischemia/reperfusion, smoking, UV irradiation, and an intake of toxic chemicals accelerate the formation of reactive oxygen species and thereafter induce the oxidation of lipids (LH), nucleic acids, and proteins. Lipid hydroperoxides, 4-hydroxynonenal, malonaldehyde, isoprostane, 8-oxoguanine, thymine glycol, protein carbonyl, nitrotyrosine, and others have been proposed as oxidative stress markers. In this review, I would like to discuss which is the sensitive marker of oxidative stress and plasma marker of oxidative stress especially in tissue.

What Is the Most Sensitive Marker of Oxidative Stress?

To answer the above question we incubated human plasma at 37°C under aerobic conditions in the presence of 5 μM cupric ion [1]. Figure 1 shows that vitamin C

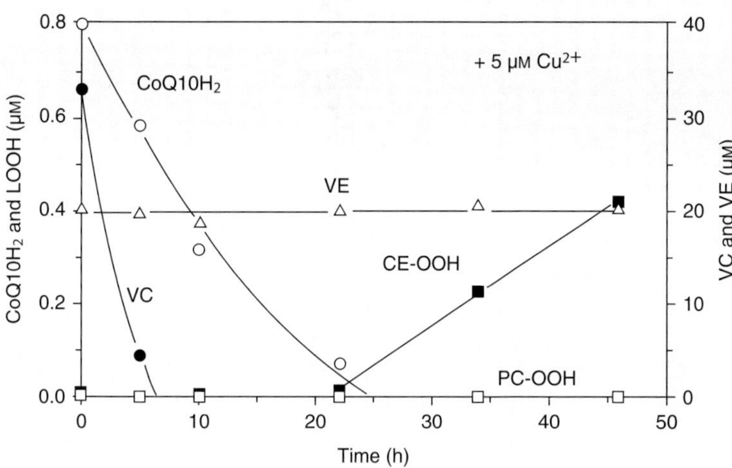

Fig. 1. Changes in plasma antioxidant levels during the aerobic oxidation of human plasma in the presence of 5 μM cupric chloride at 37°C.

(VC) decreased first and followed by the reduced form of coenzyme Q10 ($CoQ10H_2$). However, no significant decay in vitamin E (VE) was seen. A significant increase in cholesteryl ester hydroperoxides (CE-OOH) was observed after the depletion of VC and $CoQ10H_2$, indicating that VE without VC or $CoQ10H_2$ did not suppress the formation of lipid hydroperoxide.

LH are highly susceptible to oxidation and their oxidation proceeds by a free radical chain mechanism (equations 1–3). Equation 1 shows the formation of lipid radical ($L^•$) through hydrogen atom abstraction by reactive radicals ($X^•$) such as hydroxyl radical, superoxide, peroxynitrite, and others. Lipid peroxyl radical ($LOO^•$) is a key molecule for the propagation step (equations 2 and 3), and produces one molecule of lipid hydroperoxide (LOOH) per cycle. VE donates its hydrogen atom to $LOO^•$ to produce VE radical ($VE^•$), and $VE^•$ traps another $LOO^•$ to give stable products (equations 4 and 5). Thus, VE breaks lipid chain oxidation.

However, the concentration of $LOO^•$ is much smaller than that of VE in vivo. In this case, reaction 5 does not take place, and $VE^•$ abstracts hydrogen atom from lipid to give $L^•$ and VE (equation 6), although this reaction is very slow. Eventually, equations 2, 3, 4, and 6 create a new cycle where LOOH is accumulated without any loss of VE which was observed in the second half of the plasma oxidation (fig. 1). Bowry and Stocker [2] showed that VE radical acts as a chain carrier in this type of lipid oxidation, and called this oxidation process tocopherol-mediated peroxidation. Since $VE^•$ is a key molecule in the new cycle, LOOH accumulation can be blocked by reducing $VE^•$ by VC and $CoQ10H_2$ (equations 7 and 8).

$$LH + X^{\bullet} \rightarrow L^{\bullet} + XH \quad (1)$$

$$L^{\bullet} + O_2 \rightarrow LOO^{\bullet} \quad (2)$$

$$LOO^{\bullet} + LH \rightarrow LOOH + L^{\bullet} \quad (3)$$

$$LOO^{\bullet} + VE \rightarrow LOOH + VE^{\bullet} \quad (4)$$

$$VE^{\bullet} + LOO^{\bullet} \rightarrow VE\text{-}OOL \quad (5)$$

$$VE^{\bullet} + LH \rightarrow VE + L^{\bullet} \quad (6)$$

$$VE^{\bullet} + VC \rightarrow VE + VC^{\bullet} \quad (7)$$

$$VE^{\bullet} + CoQ10H_2 \rightarrow VE + CoQ10H^{\bullet} \quad (8)$$

The above results and consideration indicate that VC and $CoQ10H_2$ should be a useful marker of oxidative stress at early stage. Since $CoQ10H_2$ is converted to its oxidized form (CoQ10) by oxidation, the percentage of CoQ10 in total coenzyme Q10 (= CoQ10 + $CoQ10H_2$) (%CoQ10) would be particularly useful. Therefore, we developed a simple and reliable method for the simultaneous detection of plasma CoQ10 and $CoQ10H_2$ [3]. We applied this method to patient plasmas, and demonstrated an increase in oxidative stress in patients with hepatitis, cirrhosis, and hepatoma [4] in LEC rats (animal model of liver cancer caused by an accumulation of copper in liver) [5], in newborn babies [6], and in patients with Parkinson's disease [7]. Table 1 summarizes the representative results. Administration of copper chelator trientine to LEC rats ameliorated an increase in oxidative stress. Figure 2a shows 9 patients treated with direct percutaneous transluminal coronary angioplasty (PTCA) [8]. Plasmas were collected when hospitalized, and 0, 4, 8, 12, 16, and 20 h, and 1, 2, 3, 4, and 7 days after the PTCA. %CoQ-10 values before and right after PTCA were 9.9 ± 2.8 and 11.4 ± 2.0, respectively, reached a maximum (20–45) 1 or 2 days later, and decreased to 7.9 ± 2.7 at 7 days after PTCA, indicating an increase in oxidative stress in patients during coronary reperfusion.

Plasma Marker of Cellular Oxidative Stress

New ideas are essential for evaluating cellular oxidative damage in the blood plasma. We focused on the plasma free fatty acids (FFA) because the activity of phospholipases A_2 and A_1 is known to increase under oxidative stress, and the FFA so produced may enter the blood stream through leakage or lysis of oxidatively damaged brain cells. If this were indeed the case, we would expect a decreased content of polyunsaturated fatty acids (PUFA) such as linoleic acid (18:2), linolenic acid (18:3), arachidonic acid (20:4), and docosahexaenoic acid (22:6) in the blood plasma, since they are highly susceptible to oxidation. It has been reported that the concentrations of PUFA are reduced but those of monoenoic acids such as oleic acid (18:1) and palmitoleic acid (16:1) increase to compensate for the oxidative loss of PUFA under various conditions of oxidative stress [9]. Such changes in the plasma were observed in rats with carbon tetrachloride poisoning [10], in LEC rats [5], and patients with hepatitis, cirrhosis, and hepatoma (table 1).

Table 1. Plasma oxidative stress marker

	n	%CoQ	%PUFA	%18:1	%16:1
Fisher rat control	5	ND	38.7±0.9	22.5±0.6	5.0±0.7
Chronic CCl4 poisoning at week 16	5	ND	23.0±1.0***	26.4±2.3*	10.5±0.5***
LEC rat at week 24	6	17.0±7.9	26.8± 1.3	23.1±1.4	8.6±1.0
LEC + trientine	12	6.2±1.4*	39.7±1.5***	21.7±0.7*	3.3±0.3***
Old normal	16	6.4±3.3	24.9±5.8	27.1±4.6	2.3±1.7
Hepatitis	28	12.9±10.3**	21.2±3.9*	33.0±5.0***	4.7±1.6***
Cirrhosis	16	10.6±6.8**	20.5±3.7*	33.4±4.7***	5.5±3.4**
Hepatoma	20	18.9±11.1***	18.7±3.0***	35.6±4.0***	8.0±3.1***

* $p < 0.05$, ** $p < 0.01$ and *** $p < 0.001$, significant differences from control or normal as determined by Student's t test.

Stearoyl-CoA desaturase (SCD) is a membrane-bound enzyme that catalyzes the biosynthesis of monounsaturated fatty acids from saturated acids [11]. Thus, 18:1 and 16:1 are produced from stearic acid (18:0) and palmitic acid (16:0), respectively. SCD expression is regulated by several factors: it is induced by a fat-free, high carbohydrate diet [12], insulin [13], the peroxisome proliferator clofibrate [14], and vitamin A [15], whereas it is reduced by a high PUFA diet [16] and leptin [17]. In addition, oxidative stress is a general stimulant of SCD activity [9], since decreased levels of PUFA due to its high susceptibility to oxidation need to be compensated for by an increase in monoenoic acids to maintain membrane fluidity [11]. High SCD activity has been implicated in oxidative stress-related diseases such as diabetes, obesity, atherosclerosis, and cancer [18]. More directly, we confirmed an increase in the hepatic activity of SCD following a single administration of carbon tetrachloride or by depletion of glutathione using its biosynthetic inhibitor, buthionine sulfoximine (data not shown).

Assessment of Plasma Marker of Oxidative Stress in Rat Brain Damage Model

The free radical scavenger, 3-methyl-1-phenyl-2-pyrazolin-5-one (edaravone), has been used to treat acute brain infarction in Japan since 2001. Male Sprague-Dawley rats weighing 206–236 g (Charles River Japan, Inc.) were divided into four groups (10 animals each). Ischemia/reperfusion (I/R) rats received 2-hour middle cerebral artery (MCA) occlusion-reperfusion, and sham-operated rats were treated similarly except

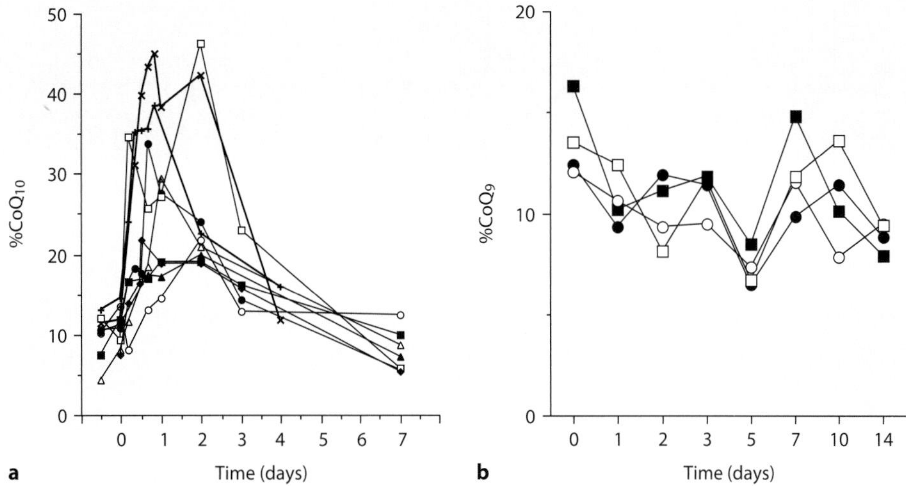

Fig. 2. Changes in plasma %CoQ-10 before and after the treatment of patients with PTCA (**a**), and changes in rat plasma %CoQ$_9$ (**b**) after 2-hour occlusion of the rat MCA and reperfusion (I/R; □) or without occlusion (sham; ■). Edaravone (3 mg/kg) was given intravenously immediately after reperfusion (single; ●) or twice a day for 2 weeks (14 days; ○).

for no I/R. Single administration group rats were infused with edaravone (3 mg/kg) intravenously immediately after the reperfusion. Finally, the same dose of edaravone was given to the rats of the repeated treatment group twice a day for 14 days [19].

Repeated treatment with edaravone significantly improved the neurological symptoms and impairment of motor function as compared to the I/R group, while single administration demonstrated limited efficacy. No significant differences in plasma antioxidants such as VC, urate, and VE, or in redox status of CoQ9 were observed among the four groups.

In contrast, the plasma %18:1 was significantly increased in the I/R group for 7 days as compared to the sham operation group (fig. 3). Oleic acid was produced from stearic acid by the action of SCD to compensate for the oxidative loss of PUFA. The above results suggest that cellular oxidative damage in the rat brain is evident for at least 7 days after I/R. Repeated treatment suppressed the %18:1 increment, while the single administration did not, which is consistent with the limited efficacy of single administration.

Plasma contents of 18:0 and 16:0 in total free fatty acids were 11–15% and 29–32%, respectively, in the male Sprague-Dawley rats used in this study. But the net increments of %18:1 and %16:1 in the I/R group were ~7 and ~2.5%, respectively. Obviously, rat SCD prefers 18:0 over 16:0 as a substrate. Therefore, %18:1 can be a more sensitive cellular oxidative damage marker than %16:1. This may be the reason why there was no significant difference in %16:1 increment between the single and repeated administration of edaravone.

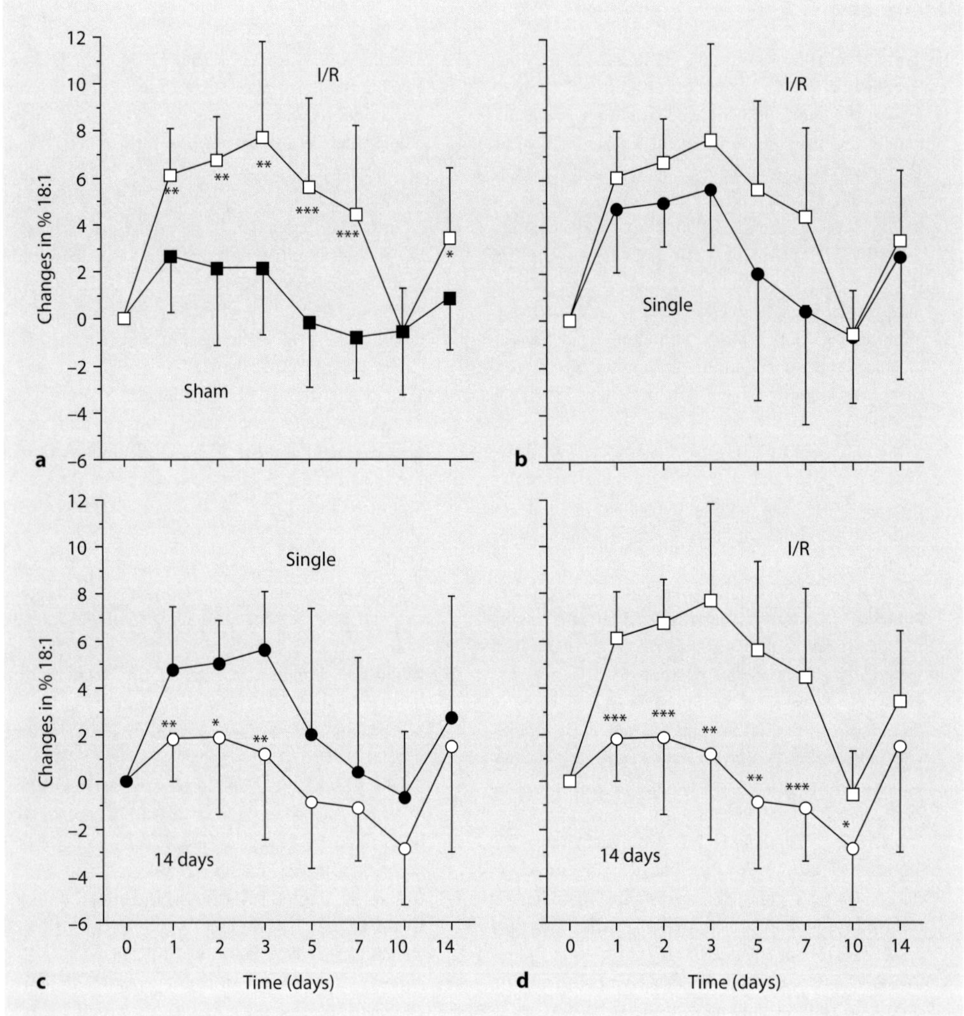

Fig. 3. Time course of changes in plasma content of oleic acid in the total free fatty acids (%18:1) following the treatments described in figure 2. Values at day 0 before occlusion were used as baseline. Points and bars show the means ± SD. * $p < 0.05$, ** $p < 0.01$ and *** $p < 0.001$, significant differences between the two groups as determined by Student's t test.

In summary, we propose that plasma %18:1 represents a good marker of brain oxidative damage, since it was increased for at least 7 days after MCA occlusion and reperfusion, but remained unchanged after the sham operation. Repeated infusion of edaravone for 2 weeks suppressed the increment of %18:1, while single administration did not. These observations are consistent with the notion that repeated treatment with edaravone could significantly improve the neurological symptoms and impairment of motor function as compared to the I/R group.

References

1 Yamamoto Y, Kawamura M, Tatsuno K, Yamashita S, Niki E, Naito C: Formation of lipid hydroperoxides in the cupric ion-induced oxidation of plasma and low density lipoprotein; in Davies KJA (ed): Oxidative Damage and Repair. New York, Pergamon Press, 1991, pp 287–291.

2 Bowry VW, Stocker R: Tocopherol-mediated peroxidation. The prooxidant effect of vitamin E on the radical-initiated oxidation of human low-density lipoprotein. J Am Chem Soc 1993;115:6029–6044.

3 Yamashita S, Yamamoto Y: Simultaneous detection of ubiquinol and ubiquinone in human plasma as a marker of oxidative stress. Anal Biochem 1997;250: 66–73.

4 Yamamoto Y, Yamashita S, Fujisawa A, Kokura S, Yoshikawa T: Oxidative stress in patients with hepatitis, cirrhosis, and hepatoma evaluated by plasma antioxidants. Biochem Biophys Res Commun, 1998; 247:166–170.

5 Yamamoto Y, Sone H, Yamashita S, Nagata Y, Niikawa H, Hara K, Nagao M: Oxidative stress in LEC rats evaluated by plasma antioxidants and free fatty acids. J Trace Elem Exp Med 1997;10:129–134.

6 Hara K, Yamashita S, Fujisawa A, Ogawa T, Yamamoto Y: Oxidative stress in newborn infants with and without asphyxia as measured by plasma antioxidants and free fatty acids. Biochem Biophys Res Commun 1999;257:244–248.

7 Sohmiya M, Tanaka M, Tak NW, Yanagisawa M, Tanino Y, Suzuki Y, Okamoto K, Yamamoto Y: Redox status of plasma coenzyme Q10 indicates elevated systemic oxidative stress in Parkinson's disease. J Neurol Sci 2004;223:161–166.

8 Yamamoto Y, Yamashita S: Plasma ubiquinone to ubiquinol ratio in patients with hepatitis, cirrhosis, hepatoma, in patients treated with percutaneous transluminal coronary reperfusion. Biofactors 1999; 9:241–246.

9 Gutteridge JMC, Quinlan GJ, Yamamoto Y: Are fatty acid patterns characteristic of essential fatty acid deficiency indicative of oxidative stress? Free Radic Res 1998;28:109–114.

10 Yamamoto Y, Nagata Y, Katsurada M, Sato S, Ohori Y: Changes in rat plasma free fatty acids composition under oxidative stress induced by carbon tetrachloride: decrease of polyunsaturated fatty acids and increase of palmitoleic acid. Redox Report 1996;2:121–125.

11 Ntambi JM: Regulation of stearoyl-CoA desaturase by polyunsaturated fatty acids and cholesterol. J Lipid Res 1999;40:1549–1558.

12 Ntambi JM: Dietary regulation of stearoyl-CoA desaturase 1 gene expression in mouse liver. J Biol Chem 1992;267:10925–10930.

13 Weiner FR, Smith PJ, Wertheimer S, Rubin CS: Regulation of gene expression by insulin and tumor necrosis factor alpha in 3T3-L1 cells: modulation of the transcription of genes encoding acyl-CoA synthetase and stearoyl-CoA desaturase-1. J Biol Chem 1991;266:23525–23528.

14 Miller CW, Ntambi JM: Peroxisome proliferators induce mouse liver stearoyl-CoA desaturase 1 gene expression. Proc Natl Acad Sci USA 1996;93:9443–9448.

15 Miller CW, Waters KM, Ntambi JM: Regulation of hepatic stearoyl-CoA desaturase gene 1 by vitamin A. Biochem Biophys Res Commun 1997;231:206–210.

16 Sessler AM, Kaur N, Palta JP, Ntambi JM: Regulation of stearoyl-CoA desaturase 1 mRNA stability by polyunsaturated fatty acids in 3T3-L1 adipocytes. J Biol Chem 1996;271:29854–29858.

17 Cohen R, Miyazaki M, Socci ND, et al : Role for stearoyl-CoA desaturase-1 in leptin-mediated weight loss. Science 2002;297:240–243.

18 Ntambi JM, Miyazaki M, Dobrzyn A: Regulation of stearoyl-CoA desaturase expression. Lipids 2004;39: 1061–1065.

19 Yamamoto Y, Yanagisawa M, Tak NW, Watanabe K, Takahashi C, Fujisawa A, Kashiba M, Tanaka M: Repeated edaravone treatment reduces oxidative cell damage in rat brain induced by middle cerebral artery occlusion. Redox Report 2009;14:251–258.

Yorihiro Yamamoto
School of Bioscience and Biotechnology, Tokyo University of Technology
1404-1 Katakura-cho
Hachioji, Tokyo 192-0982 (Japan)
Tel./Fax +81 42 637 2918, E-Mail junkan@bs.teu.ac.jp

Naito Y, Suematsu M, Yoshikawa T (eds): Free Radical Biology in Digestive Diseases.
Front Gastrointest Res. Basel, Karger, 2011, vol 29, pp 71–78

Immunochemical Detection of 4-Hydroxy-2-Nonenal-Specific Epitopes

Koji Uchida

Graduate School of Bioagricultural Sciences, Nagoya University, Nagoya, Japan

Abstract

4-Hydroxy-2-nonenal (HNE) is a major product of lipid peroxidation and is believed to be largely responsible for the cytopathological effects observed during oxidative stress. HNE exerts these effects because of its facile reactivity with biological materials, particularly proteins. Taking advantage of the fact that protein-bound HNE are excellent immunogens that are capable of stimulating adaptive immune response, a number of monoclonal antibodies against these epitopes have been developed. Moreover, using these antibodies, the HNE-specific epitopes have been detected as constituents in human patients with increased risk factors or clinical manifestations of lifestyle-related diseases, such as atherosclerosis.

Oxidative stress is increasingly seen as a major upstream component in the signaling cascade involved in many of cellular functions, such as cell proliferation, inflammatory responses, stimulating adhesion molecule, and chemoattractant production. A growing body of evidence suggests that many of the effects of cellular dysfunction under oxidative stress are mediated by products of nonenzymatic reactions, such as the peroxidative degradation of polyunsaturated fatty acids. Lipid peroxidation in tissue and in tissue fractions represents a degradative process, which is the consequence of the production and the propagation of free radical reactions primarily involving membrane polyunsaturated fatty acids, and has been implicated in the pathogenesis of numerous diseases including atherosclerosis, diabetes, cancer, and rheumatoid arthritis, as well as in drug-associated toxicity, postischemic reoxygenation injury, and aging. The peroxidative breakdown of polyunsaturated fatty acids has also been implicated in the pathogenesis of many types of liver injury and especially in the hepatic damage induced by several toxic substances.

There is increasing evidence that reactive aldehydes, such as 4-hydroxy-2-nonenal (HNE), endogenously generated during the process of lipid peroxidation are causally involved in most of the pathophysiological effects associated with oxidative stress in cells and tissues [1, 2]. More importantly, these aldehydes have been implicated as causative agents in cytotoxic processes initiated by the exposure of biological systems to oxidizing agents. Compared to free radicals, the aldehydes are relatively stable and can diffuse within or even escape from the cell and attack targets far from the site of the original event. These aldehydes can covalently react with biomacromolecules, such as protein, to generate adducts that could serve as useful biomarkers for lipid peroxidation [2]. The lipid peroxidation-specific adducts generated on protein molecules have been utilized as the biomarkers for oxidative stress. The most popular approach for the detection of lipid peroxidation-specific adducts generated on protein molecules must be the use of antibodies. Antibodies are raised by immunizing animals with carrier proteins, such as keyhole limpet hemocyanoin (KLH), that had been treated with lipid peroxidation products, and targeting specific antigenic structures (lipid peroxidation-specific epitopes) generated on the amino acid side chains of proteins. Taking advantage of the fact that the lipid peroxidation-specific epitopes are excellent immunogens that are capable of stimulating adaptive immune response, a number of monoclonal antibodies against these epitopes have been developed. Here, I provide an overview of studies on HNE-specific epitopes, focusing on their chemical structures and development of monoclonal antibodies.

4-Hydroxy-2-Nonenal-Specific Adducts

HNE, among the reactive aldehydes, is a major product of lipid peroxidation, and is believed to be largely responsible for the cytopathological effects observed during oxidative stress [1]. HNE exerts these effects because of its facile reactivity with biological materials, particularly the sulfhydryl groups of proteins. The reaction of HNE with sulfhydryl groups leads to the formation of thioether adducts that further undergo cyclization to form cyclic hemiacetals (fig. 1). Although HNE also forms Michael adducts with the imidazole moiety of histidine residues and the ε-amino group of lysine residues [2], the formation of thiol-derived Michael adducts, stabilized as the cyclic hemiacetal, is considered to constitute the main reactivity of HNE, due to the nucleophilic potential of the sulfhydryl group compared with those of the imidazole and amine groups. Balogh et al. [3] recently characterized the stereochemical configurations of the HNE-glutathione adduct by NMR experiments in combination with simulated annealing structure determinations. To differentiate between various modes of carbonyl group formation, a method for the detection and quantification of protein carbonyl groups associated with the conjugation of protein sulfhydryl groups with lipid peroxidation products was previously developed [4]. This method is based on the reduction of the adducts with $NaB[^3H]H_4$ to stable radioactive derivatives followed by cleavage of the thioether linkage upon treatment with Raney nickel. Although this procedure is not specific for the HNE-cysteine adducts, it can provide a

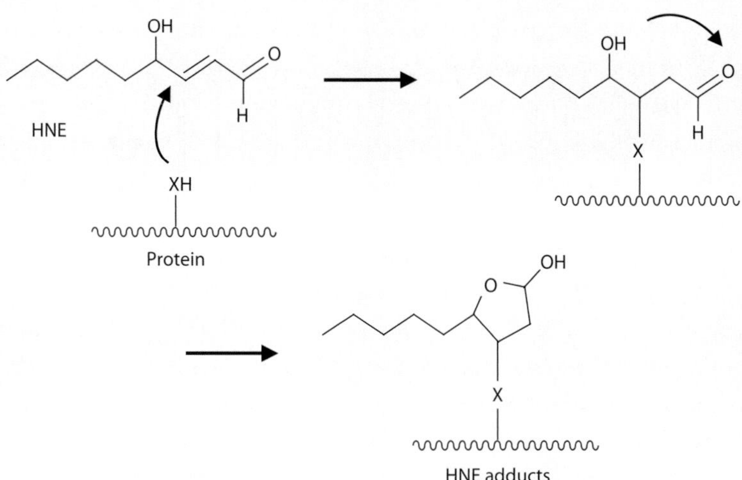

Fig. 1. Formation of hemiacetal type Michael adducts upon reaction of protein with HNE. XH represents nucleophilic amino acid side-chains (Cys, His, Lys).

means of determining the fraction of total free carbonyl groups introduced into proteins via reaction of a,b-unsaturated aldehydes with protein sulfhydryl groups.

Structural information on the nature of the HNE modification of histidine side chain was first reported by Uchida and Stadtman [5]. The HNE-histidine Michael adduct is readily isolable and is stabilized toward retro-Michael reaction, apparently on account of the poorer leaving group ability of imidazole over amine at neutral conditions. It was first speculated that the HNE-histidine adduct was a mixture of the isomeric form of the N^{π}- and N^{τ}-substituted adducts of the imidazole ring; however, on the basis of the NMR spectral analysis of the adducts, it appeared that the reaction exclusively occurs at one position (N^{τ}-alkylation). The observation that reduction of the aldehyde group of the primary Michael addition product with sodium borohydride converts them to the hydroxy derivatives that are stable to strong acid hydrolysis formed the basis of methods for the identification and quantification of the HNE-histidine Michael adduct of proteins by conventional amino acid analytical techniques [5]. It was shown by means of these techniques that at least 80% of the histidine residues that were lost when human plasma low-density lipoprotein was treated with HNE were accounted for as the Michael addition product [6].

The reaction of HNE with protein-based lysine e-amino groups has been documented in several instances [1]. In particular, it has been suggested that covalent modification of LDL by HNE may be of pathophysiologic importance. Upon reaction with lysine, HNE forms multiple products, including the HNE-lysine Michael adduct, pyrrole-type adduct, and fluorescent cross-linking-type adducts. Isolation and a proposed structure of the HNE-lysine Michael adduct were first reported by Uchida and Stadtman [7] and by Szweda et al. [8]. Later, structural definition of the HNE-lysine Michael adduct was done by Nadkarni and Sayre [9], demonstrating that the HNE-lysine Michael adduct is predominantly produced in the presence of a large excess of lysine over HNE and that the use of equimolar quantities affords mainly adducts containing two (or more) molecules of HNE per amine molecule. The pyrrole adduct is the first Schiff base-derived HNE-amine adduct [10], which is apparently very stable and represents the most stable end product of HNE protein modification. The amine-based HNE adducts which are formed rapidly are later shown to represent simple amine Michael adducts and Schiff base Michael adduct 1:2 cross-links [9]. On the other hand, Esterbauer et al. [11] demonstrated that treatment of LDL with HNE generates the same lipofuscin-like fluorescence properties as seen in oxidized LDL. This finding suggested that HNE could be a major contributor to the fluorescence generated in oxidized LDL. The chemical nature of the fluorophore arising from HNE protein modification had remained elusive; however, Itakura et al. [12] have identified for the first time the major lipofuscin-like fluorophore derived from HNE and lysine to be the 3-hydroxy-3-imino-1,2-dihydropyrrole derivative and found that the fluorescent properties of this pigment are similar to those of the oxidized LDL. Later, the same adduct was also reported by Xu and Sayre [13] and by Tsai et al. [14]. The first proposed pathway for the formation of the fluorophore was that the e-amino group of lysine reacts readily with C-1 and C-3 of HNE via Schiff base formation and Michael addition, respectively, to form the initial 1:2 HNE-amine intermediate, which is subsequently converted into the fluorophore via two oxidation steps and intermolecular cyclization. Mechanistic studies on the HNE-derived fluorophore formation have proposed an alternative mechanism, involving two 2e oxidations following initial Schiff base formation [13]. 4-Oxo-2-nonenal (ONE), which forms this fluorophore far more effectively than HNE, has been recently identified as the lipid peroxidation product [15], suggesting that ONE rather than HNE may be a potential source of this fluorophore in biological systems.

Immunochemical Detection of HNE-Specific Epitopes

Because of the increasing interest in HNE and HNE modification of proteins under oxidative stress, it seemed useful to prepare an antibody interacting specifically with the HNE moiety or with the HNE-amino acid conjugates in proteins; such antibodies have been prepared by immunizing rabbits with HNE-treated KLH [16], in which HNE adducts of histidine, lysine, and cysteine serve as the antigenic sites. Later, Toyokuni et al. [17] raised the monoclonal antibody (HNEJ2) against HNE-modified KLH, and found that the antibody cross-reacted specifically with HNE-modified proteins and had a higher affinity for HNE-histidine adduct than for HNE-lysine and HNE-cysteine adducts. This monoclonal antibody was attested to be specific for the

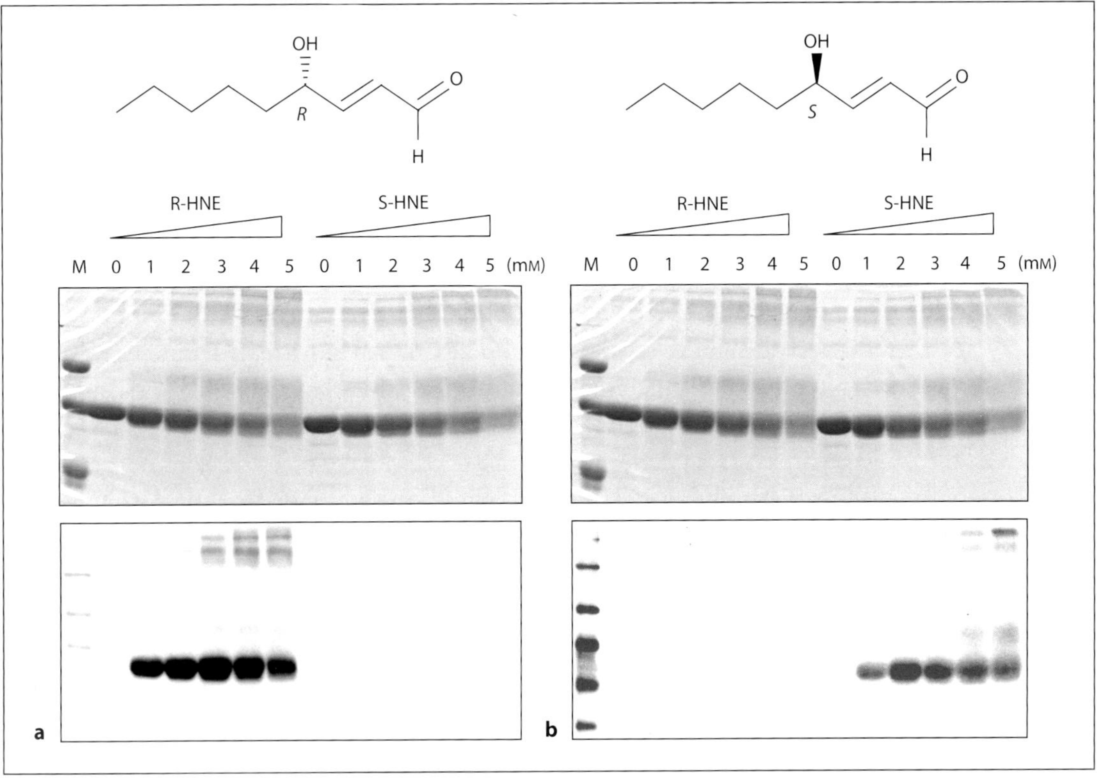

Fig. 2. Specificity of monoclonal antibodies R310 and S412 to the *R*-HNE-modified and *S*-HNE-modified proteins. **a** Immunoreactivity of R310 toward *R*-HNE-modified and *S*-HNE-modified BSA. Top: SDS-PAGE; bottom: immunoblot. **b** Immunoreactivity of S412 toward *R*-HNE-modified and *S*-HNE-modified BSA. Top: SDS-PAGE; bottom: immunoblot. **a**, **b** BSA (1 mg/ml) was incubated with *R*- or *S*-HNE (0–5 mM) in 1 ml of 50 mM sodium phosphate buffer, pH 7.4, for 24 h at 37°C.

HNE-histidine Michael adduct. In a later study, Waeg et al. [18], following independently the same line of research, reported on an antibody, which also seems to be highly specific for HNE bound to histidine residues. On the other hand, antibodies against the HNE-lysine adduct have also been raised. Sayre et al. [19] raised and epitopically characterized the specific polyclonal antibodies recognizing the lysine-based 2-pentylpyrrole adducts. The polyclonal and monoclonal antibodies against HNE-lysine fluorophore (3-hydroxy-3-imino-1,2-dihydropyrrole) have also been prepared against the fluorophore-protein conjugate [14, 20].

On the other hand, most of the antibodies, including HNEJ2, directed to the HNE-modified protein have been raised against a protein treated with racemic HNE. Therefore, nothing was known about which of the two enantiomers bound to proteins is generated in vitro and in vivo. However, Hashimoto et al. [21] successfully raised monoclonal antibodies, R310 and S412, against *R*-HNE-treated and *S*-HNE-modified KLH, respectively (fig. 2). It was observed that R310 and S412 showed the highest affinity for the *R*-HNE-treated and *S*-HNE-treated proteins, respectively, and scarcely reacted with the proteins treated with other aldehydes. The lack of cross-reactivity of the antibodies for the 2-nonenal-treated protein can be ascribed to the absence of the 4-hydroxy group which leads the primary Michael adduct to the tetrahydrofuran derivative through an intramolecular cyclization. This and the observation that both antibodies did not cross-react with the proteins that had been treated with the HNE analogs, such as 4-hydroxy-2-pentenal, 4-hydroxy-2-hexenal, 4-hydroxy-2-heptenal, 4-hydroxy-2-octenal, and 4-hydroxy-2-decenal, suggest that both tetrahydrofuran and butyl moieties of the HNE-histidine adduct may be critical for the antibody binding. In addition, the binding of these antibodies to the HNE-treated proteins was selectively inhibited by HNE-histidine adducts, suggesting that the imidazole ring is also involved in the antibody binding. Thus, we propose that R310 and S412 recognize the N^{τ}-(2-hydroxy-5-butyltetrahydrofuran-4-yl)imidazole as the common epitopes. Furthermore, the antibodies' ability to recognize the configurations of the tetrahydrofuran moiety of the HNE-histidine adduct was characterized, showing that R310 and S412 preferentially reacted with the mixture of 2*R*,4*S*,5*R* and 2*S*,4*S*,5*R* isomers and the mixture of 2*R*,4*S*,5*S* and 2*S*,4*S*,5*S* isomers, respectively. Based on the common configurations in each mixture, we suggested that both antibodies might recognize the configurations at C-4 and C-5 of the tetrahydrofuran ring in the adduct, i.e. R310 and S412 recognize the 4*S*,5*R* and 4*S*,5*S* configurations, respectively. Crystallization of R310 Fab in the presence of the HNE-histidine adduct shows that the adduct binds to a hydrophobic pocket in the groove, the antigen-binding site, consisting of six complementarity-determining regions [22]. The crystal structure of R310 Fab showed that the antibody primarily binds the HNE-histidine adduct by sandwiching between hydrophobic domains composed of several critical stacking Tyr and Phe residues. Thus, hydrophobic desolvation may underlie the observed negative enthalpy of binding.

The presence of immunoreactive materials with R310 and S412 in vivo was demonstrated in the kidney of rats exposed to Fe^{3+}-NTA. It has been suggested that oxidative stress is one of the basic mechanisms of Fe^{3+}-NTA-induced acute renal injury and is closely associated with renal carcinogenesis. The present study using R310 and S412 demonstrated that, in agreement with the previous observation on racemic HNE [17], both *R*-HNE and *S*-HNE epitopes were mainly detected in the proximal tubules, the target organs of Fe^{3+}-NTA (fig. 3). However, the intracellular localizations of these epitopes were markedly different. The *S*-HNE epitopes were detected in cytosols and in some of the nuclei, whereas the *R*-HNE epitopes were mainly detected in the nuclei. Although the mechanism of the distinct localization of *R*- and *S*-HNE epitopes is currently unknown, these results invite speculation that there may be a differential mechanism for production of *R*- and *S*-HNE or there may be specific targets of each enantiomer in the cells. Clearly, it is important to establish the mechanism for the differential cellular distributions of *R*- and *S*-HNE epitopes in the cells. Furthermore, the characterization of the biological consequences of their distributions also merits immediate attention.

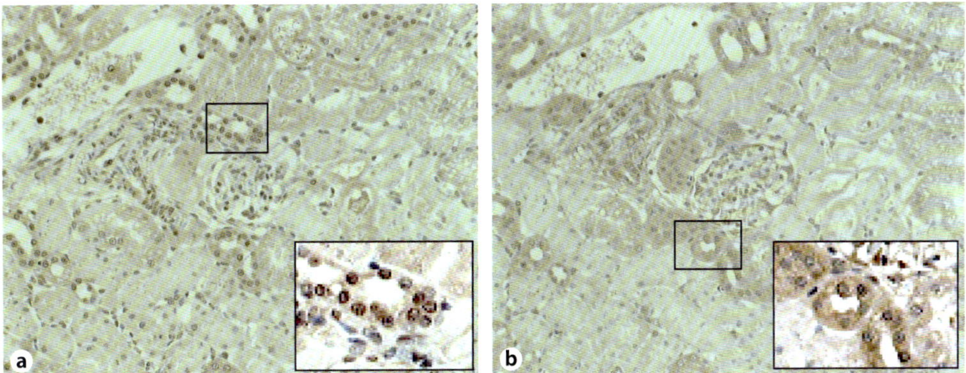

Fig. 3. Immunohistochemistry of renal cortex with R310 (**a**) and S412 (**b**). Serial sections. ×400. The immunoreactivities appeared in some of the renal proximal tubular cells 3 h after the administration of 15 mg Fe/kg body weight of Fe^{3+}-NTA. The immunoreactivites with S412 were mainly detected in the cytoplasm and in some of the nuclei, whereas the immunoreactivities with R310 were mainly detected in the nuclei.

Conclusion

On the basis of a large number of reports concerning the detection of HNE-specific adducts as biomarkers in human diseases, there is no doubt that the steady-state levels of HNE increase under pathophysiological states associated with oxidative stress. Considerable progress has also recently been made toward understanding the mechanisms of action of HNE. The quantitative and analytical importance of HNE-specific adducts has prompted the development of methods to specifically analyze these adducts because of understanding of their chemical nature, formation pathway, and distribution level in vivo. Immunological detection is a powerful tool that can be used to evaluate the presence of a desired target and its subcellular localization. The major advantages of this technique over the chemical approaches are the evaluation of small numbers of cells or archival tissues that may otherwise not be subject to analysis. The antibodies developed through this procedure bind not only to the modified protein used as the immunogen, but to a variety of other proteins on which the same epitope is found. Thus, antibodies generated against a modified protein with HNE-specific products recognize a variety of similarly modified proteins. Such properties make the antibodies directed to the epitopes useful and reliable for the immunological evaluation of lipid peroxidation in vitro and in vivo.

References

1 Esterbauer H, Schaur RJ, Zollner H: Chemistry and biochemistry of 4-hydroxynonenal, malondialdehyde and related aldehydes. Free Radic Biol Med 1991;11:81–128.

2 Uchida K: 4-Hydroxy-2-nonenal: a product and mediator of oxidative stress. Prog Lipid Res 2003;42:318–343.

3 Balogh LM, Roberts AG, Shireman LM, Greene RJ, Atkins WM: The stereochemical course of 4-hydroxy-2-nonenal metabolism by glutathione S-transferases. J Biol Chem 2008;283:16702–16710.

4 Uchida K, Stadtman ER: Modification of histidine residues in proteins by reaction with 4-hydroxynonenal. Proc Natl Acad Sci USA 1992a; 89:4544–4548.

5 Uchida K, Stadtman ER: Selective cleavage of thioether linkage in protein modified with 4-hydroxynonenal. Proc Natl Acad Sci USA 1992b; 89:5611–5615.

6 Uchida K, Szweda LI, Chae HZ, Stadtman ER: Immunochemical detection of 4-hydroxy-2-nonenal modified proteins in oxidized hepatocytes. Proc Natl Acad Sci USA 1993;90:8742–8746.

7 Uchida K, Stadtman ER: Covalent attachment of 4-hydroxynonenal to glyceraldehyde-3-phosphate dehydrogenase: a possible involvement of intramolecular and intermolecular cross-linking reactions. J Biol Chem 1993;268:6388–6393.

8 Szweda LI, Uchida K, Tsai L, Stadtman ER: Inactivation of glucose-6-phosphate dehydrogenase by 4-hydroxy-2-nonenal: selective modification of an active-site lysine. J Biol Chem 1993;268:3342–3347.

9 Nadkarni DV, Sayre LM: Structural definition of early lysine and histidine adduction chemistry of 4-hydroxynonenal. Chem Res Toxicol 1995;8:284–291.

10 Sayre LM, Arora PK, Iyer RS, Salomon RG: Pyrrole formation from 4-hydroxynonenal and primary amines. Chem Res Toxicol 1993;6:19–22.

11 Esterbauer H, Koller E, Slee RG, Koster JF: Possible involvement of the lipid-peroxidation product 4-hydroxynonenal in the formation of fluorescent chromolipids. Biochem J 1986;239:405–409.

12 Itakura K, Osawa T, Uchida K: Structure of a fluorescent compound formed from 4-hydroxy-2-nonenal and Na-hippuryllysine: a model for fluorophores derived from protein modifications by lipid peroxidation. J Org Chem 1998;63:185–187.

13 Xu G, Sayre LM: Structural characterization of a 4-hydroxy-2-alkenal-derived fluorophore that contributes to lipoperoxidation-dependent protein cross-linking in aging and degenerative disease. Chem Res Toxicol 1998;11:247–251.

14 Tsai L, Szweda PA, Vinogradova O, Szweda LI: Structural characterization and immunochemical detection of a fluorophore derived from 4-hydroxy-2-nonenal and lysine. Proc Natl Acad Sci USA 1998;95:7975–7980.

15 Lee SH, Blair IA: Characterization of 4-oxo-2-nonenal as a novel product of lipid peroxidation. Chem Res Toxicol 2000;13:698–702.

16 Uchida K, Toyokuni S, Nishikawa K, Kawakishi S, Oda H, Hiai H, Stadtman ER: Michael addition-type 4-hydroxy-2-nonenal adducts in modified low density lipoproteins: markers for atherosclerosis. Biochemistry 1994;33:12487–12494.

17 Toyokuni S, Miyake N, Hiai H, Hagiwara M, Kawakishi S, Osawa T, Uchida K: The monoclonal antibody specific for the 4-hydroxy-2-nonenal histidine adduct. FEBS Lett 1995;359:189–191.

18 Waeg G, Dimsity G, Esterbauer H: Monoclonal antibodies for detection of 4-hydroxynonenal modified proteins. Free Radic Res 1996;25:149–159.

19 Sayre LM, Sha W, Xu G, Kaur K, Nadkarni D, Subbanagounder G, Salomon RG: Immunochemical evidence supporting 2-pentylpyrrole formation on proteins exposed to 4-hydroxy-2-nonenal. Chem Res Toxicol 1996;9:1194–1201.

20 Itakura K, Oya-Ito T, Osawa T, Yamada S, Toyokuni S, Shibata N, Kobayashi M, Uchida K: Detection of lipofuscin-like fluorophore in oxidized human low-density lipoprotein. 4-hydroxy-2-nonenal as a potential source of fluorescent chromophore. FEBS Lett 2000;73:249–253.

21 Hashimoto M, Sibata T, Wasada H, Toyokuni S, Uchida K: Structural basis of protein-bound endogenous aldehydes. Chemical and immunochemical characterizations of configurational isomers of a 4-hydroxy-2-nonenal-histidine adduct. J Biol Chem 2003;278:5044–5051.

22 Akagawa M, Ito S, Toyoda K, Ishii Y, Tatsuda E, Yamaguchi S, Shibata, T, Ishino K, Kishi Y, Adachi T, Tsubata T, Takasaki Y, Hattori N, Matsuda T, Uchida K: Bispecific antibodies against modified protein and DNA with oxidized lipids. Proc Natl Acad Sci USA 2006;103:6160–6165.

Koji Uchida
Graduate School of Bioagricultural Sciences, Nagoya University
Nagoya 464-8601 (Japan)
Tel. +81 52 789 4127, Fax +81 52 789 5296, E-Mail uchidak@agr.nagoya-u.ac.jp

Naito Y, Suematsu M, Yoshikawa T (eds): Free Radical Biology in Digestive Diseases.
Front Gastrointest Res. Basel, Karger, 2011, vol 29, pp 79–84

Carbon Monoxide-Dependent Regulation of Microvascular Function: One Gas Regulates Another

Makoto Suematsu · Mayumi Kajimura · Yasuaki Kabe

Department of Biochemistry, School of Medicine, Keio University Japan Science and Technology Agency, ERATO Suematsu Gas Biology Project, Tokyo, Japan

Abstract

Carbon monoxide (CO) is a gaseous product generated by heme oxygenase (HO). Because of its nature to bind to the metal-centered prosthetic groups, the gas might bind to multiple target proteins to regulate varied biological functions. Previous studies revealed that stress-inducible levels of CO from HO-1 modulate function of different heme proteins or enzymes through binding to their prosthetic ferrous heme to regulate biological function of cells and organs including hepatobiliary systems. On the other hand, CO derived from constitutive HO-2 appeared to play house-keeping roles for microvascular perfusion or cellular homeostasis: such CO-directed target macromolecules include soluble guanylate cyclase, cytochromes P450 and NO synthase. Our recent studies revealed roles of cystathionine β-synthase in vivo for a novel CO-sensitive regulator of metabolic system that is a determinant of endogenous H_2S, another gaseous mediator. This chapter reviews the intriguing networks of metabolic regulatory mechanisms constituted by different gaseous mediators and their medical implications.

Carbon Monoxide as a Vasorelaxing Gas Discovered in the Liver

Carbon monoxide (CO) is generated by monooxygenase reaction through heme oxygenase (HO). It was first shown in the early 1970s that both nonparenchymal cells and hepatocytes are able to degrade heme to bile pigments. Kupffer cells play a role in removal and degradation of senescent erythrocytes, while hepatocytes catabolize free heme derived from hemoglobin or cytochromes P450 [1]. Distribution of HO-1 and HO-2 in the liver was previously examined in rats and humans indicating that the two isozymes have distinct topographic patterns: HO-1, the inducible form, is prominent in Kupffer cells, while the constitutive HO-2 is mostly abundant in hepatocytes [2, 3].

We showed that CO serves as a vasorelaxing factor that reduces sinusoidal tone and is necessary for maintenance of the microvascular perfusion [4, 5]. This is based on the fact that zinc protoporphyrin IX, a potent inhibitor of HO abolished CO generation and simultaneously increased the baseline vascular resistance in the perfused liver. The role of endogenous CO in sinusoidal relaxation was also shown in experiments using free oxyhemoglobin which captures both NO and CO, and methemoglobin that is able to capture NO but not CO [2]: only oxyhemoglobin was able to reproduce the vasoconstrictive effect of the HO inhibitor. Furthermore, when encapsulated with liposome, oxyhemoglobin lost its ability to constrict sinusoids because of limited accessibility to the space of Disse across endothelial fenestration, suggesting that the locus of the gas action is extrasinusoidal, perhaps involving hepatic stellate (Ito) cells. Ito cells constitute the most abundant resource of soluble guanylate cyclase in the liver that serves as a receptor for CO-responsive modest upregulation of cyclic GMP through the gas binding to the prosthetic ferrous heme. According to previous data, local concentrations of CO in the liver under physiologic conditions appeared to be around 1 μM. Under pathologic conditions including ischemia-reperfusion, acetaminophen-induced liver injury [6], endotoxemia [7] or excessive heme overloading [8], the CO concentrations reached 2–4 μM. Still unclear is the extent to which NO could interact or compete with CO in vivo, mainly because of the lack of quantitative information on functionally intact NO in situ. Although sinusoidal endothelium constitutes a major cellular component of NO synthase expression, NO released from the enzyme may be entrapped by circulating erythrocytes or cancelled by superoxide anion spontaneously released from Kupffer cells [9]. Abundant amounts of asymmetric dimethyl arginine in the normal liver tissue might be another important mechanism causing low levels of actual NO release from the enzyme resource [unpubl. obs.]. Since the potency of NO to activate the cyclase is far greater than that of CO [10], CO appears to enable to activate soluble guanylate cyclase in vivo, only when local NO concentrations are low.

Vasoconstrictive Properties of CO in Brain Microcirculation

We reported that CO acts as a tonic inhibitor of NO-mediated vasodilatation in pial microcirculation of the brain [11]. This is likely to contradict other studies reporting that CO acts as a vasodilator in other organs including the liver [12]. However, several possibilities exist that may help to explain this discrepancy. First, the principal target molecules for CO could be different in different experimental systems. It is reasonable to speculate that in the adult rat model, the major action of CO is tonic inhibition of NOS activity which results in attenuating cGMP-dependent vasodilatory mechanisms (fig. 1). Second, expression levels of CO- and NO-producing enzymes and those of target proteins could differ widely in different experimental

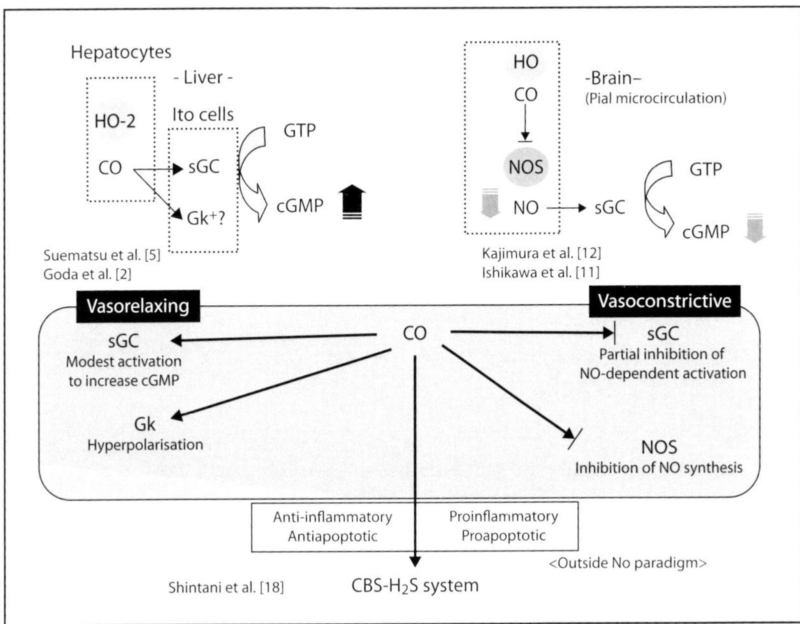

Fig. 1. Biological actions of CO and its molecular targets in vivo. sGC = Soluble guanylate cyclase; Gk = potassium channels.

models and species. For instance, in cultured cerebellar granule cells of neonatal rats, Ingi et al. [13] demonstrated that production of CO decreased with the maturation of cells, whereas that of NO increased. More importantly, in mature cultures where NO production reached a high level, HO inhibition potentiated the NO-mediated cGMP increase.

What is the physiologic implication of having differential CO effects, particularly in the brain? We postulate that the effect of CO in regulating vascular tone is multifaceted, where low tissue availability of NO renders CO a vasodilator, whereas high tissue NO availability renders CO a constrictor [14]. Under resting conditions in the rat brain, endogenously produced CO could keep blood vessels from unnecessary dilation by suppressing local NOS activity, which may, in turn, contribute to the maintenance of normal intracranial pressure. Under traumatic brain injuries such as cerebral ischemia and subarachnoid hemorrhage, CO can be overproduced, at least at some time point, due to an increase in free heme that becomes available upon the degradation of heme proteins [15] and/or by a resultant induction of HO-1 [16]. In addition, substantial alterations in NO and H_2S might occur in relation to a reduced supply of O_2, substrate availability, or altered redox states of gas-producing enzymes. Thus, aberrant actions of CO together with NO, H_2S and O_2 under these pathologic circumstances deserve further investigation.

Cystathionine β-Synthase as a CO Sensor in vivo

Several lines of biochemical evidence support the concept that cystathionine β-synthase (CBS) acts as a CO sensor. First, studies using recombinant CBS have shown that CO inhibits CBS with a K_i value of approximately 5 μM [17], which is comparable to the concentration occurring in the liver. Second, murine hepatocytes express both CO-producing HO and H_2S-producing CBS [18] with additional HO-1 induced in both hepatocytes and Kupffer cells under stress conditions. The close proximity of the enzyme distributions taken together with measured CO concentrations, and the kinetics of CBS activity led us to hypothesize that CBS acts as a CO sensor in vivo.

We demonstrated that an increase in hepatic CO content caused a global decrease in transsulfuration metabolites such as cystathionine, cysteine, and hypotaurine, suggesting that the gas inhibits the transsulfuration pathway [18]. In this study, application of metabolome analysis based on capillary electrophoresis-mass spectrometry allowed us to point out this metabolic pathway as a CO-sensitive one among huge number of the metabolic paths. The technical approach of this kind seems reasonable for mining gas-responsive macromolecules in vivo, since the gases including CO share the property to bind to metal-centered prosthetic groups of proteins; a majority of such proteins belong to enzymes in metabolic systems. Among enzymes or cofactors that regulate the transsulfuration pathway, CBS is the heme-containing enzyme that actually modulates the metabolic flux of this pathway, so far as judged from the results collected by ^{15}N-methionine flux analyses using capillary electrophoresis-mass spectrometry.

CO-overproducing livers showed a decrease in labile H_2S amount, whereas the livers of heterozygous CBS knockout mice did not show any notable decrease, suggesting that the gas inhibits the activity of CBS in vivo. To note is that administration of stress-inducible levels of CO caused a decrease in hepatic H_2S contents to stimulate HCO_3^--dependent choleresis that helps solubility of organic anions in bile. Such a CO-sensitive metabolic adaptation may play a role in quality control of bile excretion under stress or disease conditions [18]. Mechanisms by which H_2S modulates biliary excretion might involve glibenclamide-sensitive Na^+-K^+-$2Cl^-$ channels in the biliary system, although whether the gas might directly bind to the channel remains unknown.

Heme Oxygenase-2: An O_2 Sensor in Neurovascular Systems?

Generation of CO by HO depends on molecular oxygen as a substrate. Because of such a nature together with constitutive expression among varied tissues including liver, brain, carotid body Glomus cells [19] and gastrointestinal epithelium and neurons, HO-2 has been proposed to be an O_2 sensor. Based on these facts, it can be said that HO-2 plays a role in local hypoxia-induced vasodilatation in the brain. If the rate of CO production by HO-2 was more sensitive to a decrease in O_2 than the rate of NO production by NOS isozymes (i.e. the O_2 affinity of an enzyme intermediate product complex at a rate-limiting step for the HO reaction is weaker than that for NOS reaction), NOS would still be able to operate at the concentration of intracellular O_2 where HO-2 can no longer produce CO efficiently. It is worth noting that an overall K_m value for O_2, which is an important determinant of enzyme function during hypoxia, does not provide a key to the HO reaction because it consists of three oxygenation steps. To gain insight into the O_2 requirements for HO, one must find out which of the three oxygenation steps limits the rate of activity and the O_2 equilibrium constant for each step. In vitro studies demonstrated that the rate-determining step of the heme degradation is the conversion of verdoheme to the ferric biliverdin complex [20]. However, while the rate-determining step involves binding of O_2 to verdoheme, which is much slower than the binding of O_2 to the heme complex, we can only speculate on the nature of the mechanisms in vivo as the HO-2 kinetic parameters are extremely difficult to determine in vivo.

Acknowledgments

This work was supported by JST, ERATO, Suematsu Gas Biology Project, Tokyo 160-8582. Establishment of metabolome analysis was supported by Global COE Project for Human Metabolomics Systems Biology as well as by Research and Development of the Next-Generation Integrated Simulation of Living Matter, a part of the Development and Use of the Next-Generation Supercomputer Project of MEXT. The authors thank the late Prof. Emeritus Hiromasa Ishii for his continuous support and suggestions for our free radical research.

References

1 Tenhunen R, Marver HS, Schmid R: The enzymatic conversion of heme to bilirubin by microsomal heme oxygenase. Proc Natl Acad Sci USA 1968;61: 748–755.

2 Goda N, Suzuki K, Naito M, Takeoka S, Tsuchida E, Ishimura Y, Tamatani T, Suematsu M: Distribution of heme oxygenase isoforms in rat liver: topographic basis for carbon monoxide-mediated microvascular relaxation. J Clin Invest 1998;101:604–612.

3 Makino N, Suematsu M, Sugiura Y, Morikawa H, Shiomi S, Goda N, Sano T, Nimura Y, Sugimachi K, Ishimura Y: Altered expression of heme oxygenase-1 in the livers of patients with portal hypertensive diseases. Hepatology 2001;33:32–42.

4 Suematsu M, Kashiwagi S, Sano T, Goda N, Shinoda Y, Ishimura Y: Carbon monoxide as an endogenous modulator of hepatic vascular perfusion. Biochem Biophys Res Commun 1994;205:1333–1337.

5 Suematsu M, Goda N, Sano T, Kashiwagi S, Egawa T, Shinoda Y, Ishimura Y: Carbon monoxide: an endogenous modulator of sinusoidal tone in the perfused rat liver. J Clin Invest 1995;96:2431–2437.

6 Mori M, Suematsu M, Kyokane T, Sano T, Suzuki H, Yamaguchi T, Ishimura Y, Ishii H: Carbon monoxide-mediated alterations in paracellular permeability and vesicular transport in acetaminophen-treated perfused rat liver. Hepatology 1999;30: 160–168.

7 Kyokane T, Norimizu S, Taniai H, Yamaguchi T, Takeoka S, Tsuchida E, Naito M, Nimura Y, Ishimura Y, Suematsu M: Carbon monoxide from heme catabolism protects against hepatobiliary dysfunction in endotoxin-treated rat liver. Gastroenterology 2001;120:1227–1240.

8 Wakabayashi Y, Takamiya R, Mizuki A, Kyokane T, Goda N, Yamaguchi T, Takeoka S, Suematsu M, Ishimura Y: Carbon monoxide overproduced by heme oxygenase-1 causes a reduction of vascular resistance in perfused rat liver. Am J Physiol Gastrointest Liver Physiol 1999;277: G1088 –G1096.
9 Bautista AP, Spitzer JJ: Inhibition of nitric oxide formation in vivo enhances superoxide release by the perfused liver. Am J Physiol 1994;266:G783–G788.
10 Kharitonov VG, Sharma VS, Pilz RB, Magde D, Koesling D: Basis of guanylate cyclase activation by carbon monoxide. Proc Natl Acad Sci USA 1995; 92:2568–2571.
11 Ishikawa M, Kajimura M, Adachi T, Maruyama K, Makino N, Goda N, Yamaguchi T, Sekizuka E, Suematsu M: Carbon monoxide from heme oxygenase-2 is a tonic regulator against NO-dependent vasodilatation in the adult rat cerebral microcirculation. Circ Res 2005;97:e104–e114.
12 Kajimura M, Shimoyama M, Tsuyama S, Suzuki T, Kozaki S, Takenaka S, Tsubota K, Oguchi Y, Suematsu M: Visualization of gaseous monoxide reception by soluble guanylate cyclase in the rat retina. FASEB J 2003;17:506–508.
13 Ingi T, Cheng J, Ronnett GV: Carbon monoxide: an endogenous modulator of the nitric oxide-cyclic GMP signaling system. Neuron 1996;16:835–842.
14 Kajimura M, Goda N, Suematsu M: Organ design for generation and reception of CO: lessons from the liver. Antioxid Redox Signal 2002;4:633–637.
15 Chang EF, Wong RJ, Vreman HJ, Igarashi T, Galo E, Sharp FR, Stevenson DK, Noble-Haeusslein LJ: Heme oxygenase-2 protects against lipid peroxidation-mediated cell loss and impaired motor recovery after traumatic brain injury. J Neurosci 2003; 23:3689–3696.
16 Suzuki H, Kanamaru K, Tsunoda H, Inada H, Kuroki M, Sun H, Waga S, Tanaka T: Heme oxygenase-1 gene induction as an intrinsic regulation against delayed cerebral vasospasm in rats. J Clin Invest 1999;104:59–66.
17 Taoka S, Banerjee R: Characterization of NO binding to human cystathionine β-synthase: possible implications of the effects of CO and NO binding to the human enzyme. J Inorg Biochem 2001;87:245–251.
18 Shintani T, Iwabuchi T, Soga T, Kato Y, Yamamoto T, Takano N, Hishiki T, Ueno Y, Ikeda S, Sakuragawa T, Ishikawa K, Goda N, Kitagawa Y, Kajimura M, Matsumoto K, Suematsu M: Cystathionine β-synthase as a carbon monoxide-sensitive regulator of bile excretion. Hepatology 2009;49:141–150.
19 Prabhakar NR, Dinerman JL, Agani FH, Snyder SH: Carbon monoxide: a role in carotid body chemoreception. Proc Natl Acad Sci USA 1995;92:1994–1997.
20 Matsui T, Nakajima A, Fujii H, Matera KM, Migita CT, Yoshida T, Ikeda-Saito M: O_2^- and $H_2O_2^-$ dependent verdoheme degradation by heme oxygenase: reaction mechanisms and potential physiological roles of the dual pathway degradation. J Biol Chem 2005;280:36833–36840.

Makoto Suematsu, MD, PhD
Department of Biochemistry, School of Medicine, Keio University
35 Shinanomachi, Shinjuku-ku
Tokyo 160-8582 (Japan)
Tel. +81 3 5363 3753, Fax +81 3 5363 3466, E-Mail msuem@sc.itc.keio.ac.jp

Naito Y, Suematsu M, Yoshikawa T (eds): Free Radical Biology in Digestive Diseases.
Front Gastrointest Res. Basel, Karger, 2011, vol 29, pp 85–96

Nrf2-Keap1 Signaling as a Common Chemopreventive Target of Antioxidative and Anti-Inflammatory Phytochemicals

Young-Joon Surh

Department of Molecular Medicine and Biopharmaceutical Sciences, Graduate School of Convergence Science and Technology and College of Pharmacy, Seoul National University, Seoul, South Korea

Abstract

A precise and timely regulation of redox balance is essential for the cellular homeostatic control. Nuclear factor erythroid derived 2-related factor-2 (Nrf2) is a transcription factor that mediates cellular defense mechanism against oxidative stress by neutralizing electrophiles or reactive oxygen species. However, recent studies have revealed that Nrf2 also has an anti-inflammatory function and hence represents an important therapeutic target for the inflammatory disorders. Thus, disruption of cellular defense response by Nrf2 deficiency has been found to cause enhanced susceptibility to not only oxidative stress but also inflammatory injuries. Although it has not been clearly elucidated, the antioxidative function of genes targeted by Nrf2 may cooperatively regulate the innate immune response and also repress the expression of proinflammatory genes. Many dietary phytonutrients can induce Nrf2-mediated cytoprotective gene expression, thereby fortifying cellular defense against oxidative and inflammatory insult. Therefore, the maintenance of cellular defense capacity through activation of Nrf2 represents an important mechanism underlying chemoprevention of oxidative stress- and inflammation-associated carcinogenesis.

Cancer Prevention with Dietary Phytochemicals

'The war against cancer is far from over. Observed changes in mortality due to cancer primarily reflect changes in incidence or early detection. The effect of new treatments for cancer on mortality has been largely disappointing ...' This statement is quoted from a seminal article published in *New England Journal of Medicine* in 1997 [1], approximately 25 years after President Nixon declared the war on cancer by signing the National Cancer Act. Even since then, despite our enormous efforts to conquer cancer, we have been losing the war on this dreadful disease. Although there has been tremendous progress in the development of chemotherapy and radiation therapy as well as surgical techniques, cancer still remains a major global health burden. By admitting the reality that there is no arsenal that can completely and selectively destroy the majority of human malignancies, we obviously need a paradigm shift for the management of cancer. Like the many other human disorders, cancer, in general, is preventable. So, one of the most promising approaches towards the control of cancer is a national commitment to prevention, with a concomitant rebalancing of the focus and funding of research [1].

One of the attractive and practical ways to reduce the risk of cancer is chemoprevention. Chemoprevention is an attempt to use of relatively safe chemical substances of either natural or synthetic origin to inhibit, retard or reverse the progress of carcinogenesis before the malignancy manifests. In fact, the promising results from numerous preclinical and limited clinical studies highlight chemoprevention as a rational strategy to fight cancer.

According to the report from the World Cancer Research Fund, about 30–40% of cancers can be prevented by appropriate food and nutrition, physical activity and avoidance of obesity [2]. Epidemiological studies also support the notion that the regular consumption of nonnutritive ingredients derived from plant-based diet, collectively termed 'phytochemicals', can reduce the risk of certain cancers [2]. A vast variety of substances present in fruits, vegetables, grains and spices have been reported to intervene in or halt a specific stage of the carcinogenic process [3]. It is now estimated that more than 1,000 different food-derived phytochemicals possess chemopreventive activities. These include resveratrol from grapes, curcumin from turmeric, epigallocatechin gallate from green tea, sulforaphane from broccoli, genistein from soybean, indole-3-carbinol from cabbage, sulforaphane from broccoli, lycopene from tomato, organosulfur compounds from garlic, gingerol from ginger, caffeic acid phenethyl ester from honey bee propolis, etc. [3].

The Intracellular Signal Network as a Prime Target of Chemopreventive Phytochemicals

In normal cells, numerous molecules and events in the cytoplasm relay signals recognized by the distinct membrane receptor to the gene expression machinery in the nucleus. Research directed toward elucidating molecular mechanisms underlying transformation of normal cells to precancerous and eventually malignant ones has identified several essential components of intracellular signal transduction pathways that are often improperly functioning during carcinogenic processes. Among these, of particular interest is a panel of redox-sensitive transcription factors. These transcription factors play a pivotal role in fine tuning of expression of a wide array of genes involved in the maintenance of homeostatic cell growth and proliferation as well as the protection of cells from oxidative and other noxious insults. Since the cellular signaling mediated by many redox-sensitive transcription factors and their target gene products often goes awry in carcinogenesis, it is fairly rational to target them for achieving chemoprevention [4, 5]. Oxidative stress and inflammatory injuries are two major culprits that are closely linked in the process of multi-stage carcinogenesis [6]. Thus, proper regulation of abnormally activated redox and inflammation signaling is important for chemoprevention as well as chemotherapy.

The generation of excessive reactive oxygen species (ROS) as byproducts of aerobic metabolism and a concomitant fall in the intrinsic antioxidant capacity of cells can lead to a state of oxidative stress, which contributes to carcinogenesis. ROS, such as superoxide anion radical, hydroperoxyl radical, hydrogen peroxide, and hydroxyl radical, contribute to tumorigenesis either directly by damaging DNA or indirectly by modulating cellular signal transduction pathways [7]. Moreover, accumulation of ROS in vivo leads to a state of persistent local inflammation. Like oxidative stress, inflammation plays a role in multistage carcinogenesis through several distinct mechanisms including direct damage to genomic DNA and alteration of intracellular signal transduction leading to abnormal cellular growth [7]. Thus, both oxidative stress and inflammation not only can initiate tumorigenesis but also promote the proliferation of damaged cells, creating a tumor microenvironment which facilitates the neoplastic transformation of premalignant cells [7].

As oxidative stress and inflammatory injuries are closely linked to each other in the process of multi-stage carcinogenesis, compounds with anti-inflammatory activities are anticipated to inhibit oxidative stress, and vice versa. We have investigated chemopreventive effects of some edible phytochemicals, with special focus on those with antioxidative and anti-inflammatory properties, and their underlying molecular mechanisms [3]. Our research program has attempted to unravel common events mediated by one distinct redox-sensitive transcription factor, nuclear factor erythroid2-related factor-2 (Nrf2), which mediates both antioxidant and anti-inflammatory signaling.

Nrf2 as an Essential Component of Cellular Antioxidant and Anti-Inflammatory Signaling

Living in an environment of various known and unknown sources of ROS, our body has the intrinsic ability to guard against oxidative stress-induced cellular damage. The induction of antioxidant enzymes and other cytoprotective proteins represents one of the most important components of cellular defense mechanisms, whereby a diverse array of oxidative toxicants and other reactive species can be eliminated from the cell before they damage genomic DNA. Naturally, cells/tissues are empowered with a panel of antioxidant and detoxifying enzymes such as NAD(P)H:quinone oxidoreductase-1 (NQO1), superoxide dismutase (SOD), glutathione *S*-transferase (GST), glutathione peroxidase (GPx), heme oxygenase-1 (HO-1), glutamate-cysteine ligase (GCL), uridine diphosphate glucuronosyltransferase (UGT), etc., which are responsible for

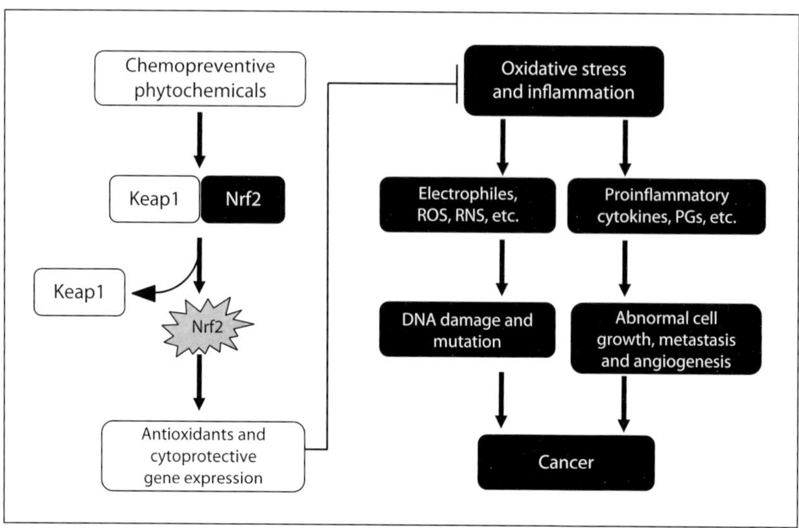

Fig. 1. Inhibition of oxidative stress- and inflammation-induced carcinogenesis by chemopreventive phytochemicals through activation of Nrf2 signaling. Chemopreventive phytochemicals facilitate elimination of electrophilic carcinogens or inactivation of other reactive species (e.g. ROS and reactive nitrogen species, RNS) through activation of Nrf2-ARE signaling and subsequent induction of detoxification and cytoprotective gene expression. Alternatively, Nrf2-driven gene expression by dietary phytochemicals can suppress excessive cell proliferation, metastasis and angiogenic processes. PGs = Prostaglandins.

protecting DNA and other critical biomolecules from damage that can be caused by ROS and electrophilic carcinogens.

The proximal promoter regions of aforementioned antioxidant and detoxification genes contain a consensus sequence known as antioxidant response element (ARE) or electrophile response element (EpRE), which is the preferred site of Nrf2 binding. The induction of a battery of antioxidant and phase II detoxifying enzymes is primarily regulated by Nrf2 as evidenced by negation of their expression in the genetically engineered Nrf2-deficient mice [8]. In addition to protection against oxidative stress, many recent studies address that Nrf2 responds to proinflammatory stimuli and rescues cells/tissues from inflammatory injuries [9]. Disruption of cellular defense response by Nrf2 deficiency causes enhanced susceptibility to both oxidative stress- and inflammation-associated disorders including cancer [9].

Under normal physiological conditions, a cytoskeleton-binding protein called Kelch-like erythroid CNC homologue-associated protein 1 (Keap1) binds to Nrf2, thereby repressing its activation. In response to oxidative or electrophilic insults, Nrf2 is dissociated from the inhibitory protein Keap1 and translocates to nucleus and binds to ARE/EpRE, leading to the coordinated upregulation of downstream genes (fig. 1) that boost cellular cytoprotective potential. Likewise, certain chemopreventive and cytoprotective phytochemicals diminish the affinity of Keap1 for Nrf2, which allows

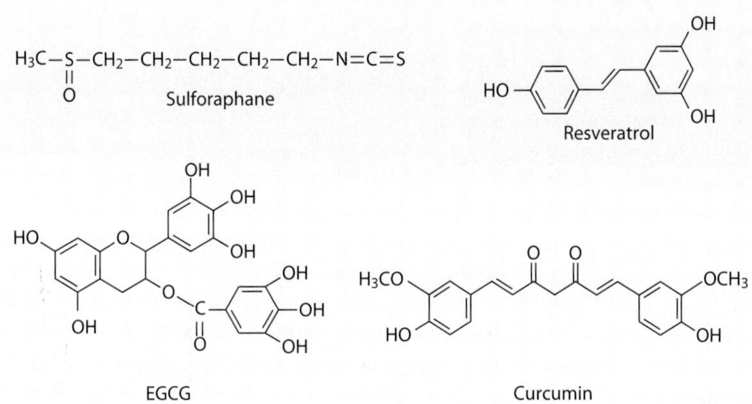

Fig. 2. Chemical structures of some representative chemopreventive phytochemicals that can activate Nrf2 signaling. Some chemopreventive phytochemicals (e.g. sulforaphane and curcumin) can directly interact with sensor cysteine thiol(s) present in the Keap1, thereby diminishing the affinity of this inhibitory protein for Nrf2. This will facilitate the dissociation of Nrf2 and its nuclear translocation, binding to ARE. Other chemicals, such as resveratrol and EGCG, activate the upstream kinases, such as PI3K and MAP kinases, inducing phosphorylation of Nrf2, which also stimulates the nuclear translocation and transcriptional activity of Nrf2.

for the translocation of Nrf2 into the nucleus [10]. Thus, Nrf2-Keap1 signaling is shared by mild stresses and some phytochemicals. While activation of Nrf2 signaling is a physiologically important stress response, the same defense mechanism is activated by some chemopreventive or cytoprotective phytochemicals (fig. 1), many of which are originally produced by plants as toxins (phytoalexins). From an evolutionary perspective, the noxious properties of phytoalexins are important for protecting plants from insects, pests, and adverse environmental conditions including oxidative stress caused by solar irradiation. However, at the subtoxic doses ingested by humans that consume the plants, these same phytochemicals may generate mild stress, thereby inducing adaptive survival responses mediated by stress-response signaling molecules, including Nrf2 [11]. In line with this notion, some chemopreventive phytochemicals capable of inducing Nrf2-driven antioxidant gene expression can act as pro-oxidants. In this context, it is noticeable that there is a good correlation between EpRE-activating activity of some flavonoids and their pro-oxidant property [12].

Chemopreventive Phytochemicals That Target Nrf2-Keap1 Signaling

Many chemopreventive substances derived from edible plants, by activating Nrf2 signaling, potentiate cellular antioxidant and anti-inflammatory defense [10]. Some representative substances (fig. 2) present in the plant-based diet that exert chemopreventive effects through Nrf2 activation are listed below.

Sulforaphane

Sulforaphane [1-isothiocyanato-4-(methylsulfinyl)butane], a chemopreventive isocyanate derived from broccoli sprouts and mature broccoli, has been reported to be a potent inducer of antioxidant and phase II carcinogen-detoxifying enzymes predominantly by activating Nrf2 [13]. The compound induced marked expression of NQO1 and GCL in the small intestine of *Nrf2* wild-type mice, while the *Nrf2*-null mice displayed lower levels of these enzymes upon sulforaphane treatment [14]. Topical application of sulforaphane induced HO-1 expression in the epidermis and inhibited chemically induced skin carcinogenesis in C57BL/6 mice harboring wild type Nrf2, whilst no such chemopreventive effects were observed in the Nrf2-deficient mice [15]. When treated to human keratinocytes in culture, sulforaphane induced Nrf2-dependent gene expression. Nrf2 activation by sulforaphane resulted in an increase in the levels of NQO1, HO-1, and GCL and their mRNA transcripts. Furthermore, sulforaphane elevated cellular reduced glutathione (GSH) levels and antagonized tumor necrosis factor-α (TNF-α)-induced NF-κB transactivation [16].

Oral administration of sulforaphane-rich broccoli sprouts to C57BL/6 female mice infected with *Helicobacter pylori* Sydney strain 1 and maintained on a high-salt (7.5% NaCl) diet reduced gastric bacterial colonization, attenuated mucosal expression of TNF-α and interleukin-1β (IL-1β), mitigated corpus inflammation, and prevented gastric corpus atrophy [17]. However, these anti-inflammatory effects were not observed in $Nrf2^{-/-}$ mice, corroborating the importance of Nrf2 activation in sulforaphane-mediated protection against gastric inflammation. Moreover, daily administration of sulforaphane-rich broccoli sprouts to mice as well as human volunteers led to decreased colonization of *H. pylori* as compared to placebo controls [17].

Pretreatment of $Nrf2^{+/+}$ primary peritoneal macrophages with sulforaphane induced HO-1 expression, whilst the compound also inhibited lipopolysaccharide-induced expression or production of TNF-α, IL-1β, cyclooxygenase-2 and inducible nitric oxide synthase [18]. The anti-inflammatory effects of sulforaphane were attenuated in $Nrf2^{-/-}$ primary peritoneal macrophages. Mechanistically, sulforaphane activated Nrf2 through enhanced phosphorylation of upstream Akt and extracellular signal-regulated protein kinase, the blockade of p38 mitogen-activated protein (MAP) kinase, and direct modification of specific cysteine residues on Keap1 [10].

Among these, of particular interest is modification of critical cysteine residues of Keap1 [19], which lead to stabilization of Nrf2 to activate the ARE of phase 2 enzymes.

Resveratrol

Resveratrol has been reported to elevate the expression and/or the activity of several antioxidants and enzymes and other related proteins involved in cytoprotection and stress response [20–24]. The compound restored cigarette smoke extract-induced depletion of cellular GSH by inducing Nrf2-driven upregulation of GCL expression in human primary small airway epithelial cells and human alveolar epithelial (A549) cells, thereby protecting these cells from cigarette smoke extract-induced oxidative damage [25]. In another study conducted with cultured human bronchial epithelial cells (HBE1), resveratrol increased the mRNA levels of GCL, NQO1, and HO-1 [26]. While resveratrol induced antioxidant or cytoprotective gene expression in HBE1 cells, the compound did not rescue the cells from cigarette smoke-induced apoptosis but rather exacerbated it [26]. Resveratrol activated Nrf2 and augmented the activities of antioxidant enzymes, including catalase, SOD, GR, GPx, NQO1 and GST, in primary rat hepatocytes challenged with oxidative stress [27]. Moreover, resveratrol increased the phosphorylation and nuclear translocation of Nrf2, and induced the activity as well as the expression of NQO1 at both protein and mRNA levels in human leukemia K562 cells [28].

Dietary administration of resveratrol (50–300 mg/kg) abrogated the oxidative stress and inflammatory response caused by the hepatocarcinogen diethylnitrosamine in rats, which appears to be associated with its upregulation of Nrf2 [29]. The above findings may account for the previously reported chemopreventive effects of resveratrol on diethylnitrosamine-induced hepatocarcinogenesis [29].

(–)-Epigallocatechi-3-Gallate

Intraperitoneal administration of (–)-epigallocatechim-3-gallate (EGCG), at a dose equivalent to four cups of 2% tea for 15 days, elevated the levels of GST, GPx, SOD and catalase in mouse liver, and reduced lipid peroxidation and cell proliferation during 7,12 dimethylbenz[*a*]anthracene-initiated and 12-*O*-tetradecanoylphorbol-13-acetate-promoted mouse skin tumorigenesis [30]. EGCG treatment inhibited the growth and metastasis of colon tumor implanted orthotopically in the cecum of nude mice [31]. In addition, the protein level of Nrf2 and the mRNA levels of Nrf2, UGT1A, UGT1A8 and UGT1A10 were significantly elevated in EGCG-treated mice in comparison with those in the control group. EGCG, given by gavage, significantly decreased 2-amino-3-methylimidazo[4,5-*f*]quinoline-induced atypical hyperplasia, the number of aberrant crypt foci and adenocarcinoma formation by activating Nrf2 and upregulating UGT1A10 in mouse colon [31]. EGCG upregulated the expression of Nrf2 and increased the level of UGT1A in the human colon cancer Caco-2 cell line [32]. Administration of EGCG (5, 10, and 20 mg/kg) by gavage on every other day up to 4 weeks resulted in enhanced mRNA and protein levels of Nrf2 and UGT1A [32]. EGCG treatment induced nuclear translocation of Nrf2 and HO-1 expression in B-lymphoblasts [33]. The EGCG-induced HO-1 expression was completely suppressed by the phosphatidylinositol 3-kinase (PI3K) inhibitor, LY294002. EGCG induced HO-1 upregulation as well as nuclear translocation of Nrf2 effectively in wild-type mouse embryo fibroblasts (MEFs), but not in Nrf2$^{-/-}$ MEFs [33]. Likewise, EGCG also enhanced the mRNA expression of HO-1 and MnSOD, and the nuclear translocation and ARE binding of Nrf2 in human mammary epithelial cells [34]. The pharmacologic inhibition of these kinases abrogated the nuclear translocation of Nrf2 induced by EGCG. EGCG ameliorated bleomycin-induced pulmonary fibrosis and restored the antioxidant status, possibly through activation of Nrf2 signaling [35]. EGCG induced HO-1 expression through Nrf2 activation in rat neurons in culture, which conferred protection against oxidative stress [36].

Global gene expression profiles in the liver and the small intestine were compared in C57BL/6J and C57BL/6J/Nrf2-deficient mice following treatment with EGCG [37]. Among the defined genes, 671 and 228 Nrf2-dependent genes were identified in the liver and the small intestine, respectively. Based on their biological functions, these genes were mainly classified into the category of ubiquitination and proteolysis, electron transport, detoxification, transport, cell growth and apoptosis, cell adhesion, kinase and phosphatases, and transcription factors [37].

Curcumin

The induction of antioxidant or detoxifying enzymes by curcumin is also mediated via the Nrf2-ARE signaling as assessed in different cell culture systems. Treatment of porcine renal epithelial cells with curcumin caused the expression of Nrf2 and a significant increase in expression of HO-1 [38]. Curcumin transiently induced expression of HO-1 and its mRNA transcript in two breast cell lines (HBL100 and MDA-MB468) [39]. The curcumin-induced HO-1 expression was abrogated by the NF-κB-DNA binding inhibitor helenalin. Curcumin also induced nuclear translocation of Nrf2 in the same cells, which was attenuated by pharmacologic inhibition of the PI3K inhibitor or the p38 inhibitor [33]. Curcumin induced expression of HO-1 and the regulatory subunit of GCL (GCLM) and Nrf2 binding to the ARE in human monocytes [39]. Rottlerin (a protein kinase Cδ inhibitor) and protein kinase Cδ antisense oligonucleotides significantly inhibited curcumin-induced GCLM and HO-1 mRNA expression and Nrf2 ARE binding [39]. Curcumin treatment caused ROS generation, which contributed to activation of Nrf2 and p38 MAP kinase and subsequently HO-1 expression in HUH7 human hepatoma cells [40]. Curcumin also induced the expression of human GST P1 in human hepatoma HepG2 cells [41]. Curcumin was found to induce HO-1 expression, presumably through Nrf2-dependent ARE activation, in rat vascular smooth muscle cells and human aortic smooth muscle cells, which accounts for the antiproliferative effect of curcumin [42]. Tetrahydrocurcumin, unable to act as a Michael reaction acceptor due to lack of the α,β-unsaturated carbonyl moiety, failed to induce HO-1 expression, ARE activation and inhibition of vascular smooth muscle cell growth [42]. It was reported that the activation of PI3K and p38 MAP kinase by curcumin augmented the expression of the aldose reductase gene via Nrf2 in vascular smooth muscle cells, and that increased aldose reductase activity may represent an important cellular adaptive response against oxidative stress [43]. Dietary administration of curcumin elevated the Nrf2 protein levels and enhanced its nuclear translocation and increased DNA binding in the liver and the lung of Swiss albino mice [44]. According to this study, elevated protein and mRNA levels and the activities of hepatic GST and NQO1 in mice given curcumin are considered to facilitate the detoxification of benzo[*a*]pyrene [44]. In agreement with the observed curcumin-mediated induction of carcinogen detoxifying enzymes, there was concomitant reduction in benzo[*a*]pyrene-induced DNA adduct formation, oxidative damage and inflammation. Oral administration of curcumin, but not its nonelectrophilic analog tetrahydrocurcumin, also enhanced the nuclear translocation and the ARE binding of Nrf2, and induced the expression of HO-1 in the liver of male ICR mice, protecting the animals against dimethylnitrosamine-induced hepatotoxicity [45].

Pretreatment of retina-derived cell lines (661W and ARPE-19) with curcumin protected these cells from H_2O_2-induced oxidative damage by upregulating cellular antioxidant proteins, such as HO-1 and thioredoxin [46]. Curcumin also induced Nrf2 protein expression with upregulation of GCL mRNA, and increased the levels of cellular antioxidant GSH in iKKU-M214 CCA cells [47].

Conclusion

Nrf2-mediated induction of cytoprotective gene expression is important not only in elimination or neutralization of ROS and electrophilic toxicants, but also blocking deleterious production of proinflammatory mediators. Nrf2 was shown to inhibit the gene expression of inflammatory factors, such as cytokines, chemokines, cell adhesion molecules, and cyclooxygenase-2. Consistent with these findings, Nrf2 deficiency causes exacerbated inflammatory as well as oxidative injuries. Thus, the coordinated induction of genes targeted by Nrf2 plays a pivotal role in maintaining the cellular redox balance and anti-inflammatory responses. As chronic inflammation and excessive oxidative stress are two major culprits in the majority of human cancers, Nrf2 represents one of the most critical targets in identifying promising chemopreventive agents, including dietary phytochemicals, and exploring their underlying molecular mechanisms.

Acknowledgements

This study was supported by the research grant from the Korea Science and Engineering Foundation for Biofoods Research Program, Ministry of Education, Science and Technology, Republic of Korea.

References

1 Suit H, Isselbacher K, Chabner B: Winning the war on cancer. N Eng J Med 1997;337:936–937.

2 WCRF/American Institute of Cancer Research: Food, Nutrition, Physical Activity and the Prevention of Cancer: A Global Perspective. Washington, American Institute of Cancer Research, 2007, pp 75–114.

3 Surh Y-J: Cancer chemoprevention with dietary phytochemicals. Nat Rev Cancer 2003;3:768–780.

4 Kundu JK, Surh YJ: Molecular basis of chemoprevention with dietary phytochemicals: redox-regulated transcription factors as relevant targets. Phytochem Rev 2009;8:333–347.

5 Surh Y-J, Kundu JK, Na HK, Lee JS: Redox-sensitive transcription factors as prime targets for chemoprevention with anti-inflammatory and antioxidative phytochemicals. J Nutr 2005;135:2993S–3001S.

6 Federico A, Morgillo F, Tuccillo C, Ciardiello F, Loguercio C: Chronic inflammation and oxidative stress in human carcinogenesis. Int J Cancer 2007; 121:2381–2386.

7 Kundu JK, Surh Y-J: Inflammation: gearing the journey to cancer. Mutat Res 2008;659:15–30.

8 Kwan MK, Kensler T: Targeting NRF2 signaling for cancer chemoprevention. Toxicol Appl Pharmacol 2010;244:66–76.

9 Kim JY, Cha YN, Surh Y-J: A protective role of nuclear factor-erythroid 2-related factor-2 (Nrf2) in inflammatory disorders. Mutat Res 2010;690:12–23.

10 Surh Y-J, Kundu JK, Na HK: Nrf2 as a master redox switch in turning on the cellular signaling involved in the induction of cytoprotective genes by some chemopreventive phytochemicals. Planta Med 2008; 74:1526–1539.

11 Son TG, Camandola S, Mattson MP: Hormetic dietary phytochemicals. Neuromol Med 2008;10: 236–246.

12 Lee-Hilz YY, Boerboom A-MJF, Westphal AH, van Berkel WJH, Aarts JMMJG, Rietjens IMCM: Prooxidant activity of flavonoids induces EpRE-mediated gene expression. Chem Res Toxicol 2006; 19:1499–1505.

13 Juge N, Mithen RF, Traka M: Molecular basis for chemoprevention by sulforaphane: a comprehensive review. Cell Mol Life Sci 2007;64:1105–1127.

14 Thimmulappa RK, Mai KH, Srisuma S, Kensler TW, Yamamoto M, Biswal S: Identification of Nrf2-regulated genes induced by the chemopreventive agent sulforaphane by oligonucleotide microarray. Cancer Res 2002;62:5196–5203.

15 Xu C, Huang MT, Shen G, Yuan X, Lin, W, Khor TO, Conney AH, Kong AN: Inhibition of 7,12-dimethylbenz(*a*)anthracene-induced skin tumorigenesis in C57BL/6 mice by sulforaphane is mediated by nuclear factor E2-related factor 2. Cancer Res 2006;66:8293–8296.

16 Wagner AE, Ernst I, Iori R, Desel C, Rimbach G: Sulforaphane but not ascorbigen, indole-3-carbinole and ascorbic acid activates the transcription factor Nrf2 and induces phase-2 and antioxidant enzymes in human keratinocytes in culture. Exp Dermatol 2010;19:137–144.

17 Yanaka A, Fahey JW, Fukumoto A, Nakayama M, Inoue S, Zhang S, Tauchi M, Suzuki H, Hyodo I, Yamamoto M: Dietary sulforaphane-rich broccoli sprouts reduce colonization and attenuate gastritis in *Helicobacter pylori*-infected mice and humans. Cancer Prev Res (Phila Pa) 2009;2:353–360.

18 Lin W, Wu RT, Wu T, Khor TO, Wang H, Kong AN: Sulforaphane suppressed LPS-induced inflammation in mouse peritoneal macrophages through Nrf2 dependent pathway. Biochem Pharmacol 2008; 76:967–973.

19 Hong F, Freeman ML, Liebler DC: Identification of sensor cysteines in human Keap1 modified by the cancer chemopreventive agent sulforaphane. Chem Res Toxicol 2005;18:1917–1926.

20 Chen CY, Jang J-H, Li MH, Surh Y-H: Resveratrol upregulates heme oxygenase-1 expression via activation of NF-E2-related factor 2 in PC12 cells. Biochem Biophys Res Commun 2005;331:993–1000.

21 Calabrese V, Cornelius C, Mancuso C, Pennisi G, Calafato S, Bellia F, Bates TE, Giuffrida Stella AM, Schapira T, Dinkova Kostova AT, Rizzarelli E: Cellular stress response: a novel target for chemoprevention and nutritional neuroprotection in aging, neurodegenerative disorders and longevity. Neurochem Res 2008;33:2444–2471.

22 Vincent AM, Kato K, McLean LL, Soules ME, Feldman EL: Sensory neurons and schwann cells respond to oxidative stress by increasing antioxidant defense mechanisms. Antioxid Redox Signal 2009;11:425–438.

23 Ungvari Z, Bagi Z, Feher A, Recchia FA, Sonntag WE, Pearson K, de Cabo R, Csiszar A: Resveratrol confers endothelial protection via activation of the antioxidant transcription factor Nrf2. Am J Physiol Hear Circ Physiol 2010;299:H18–H24.

24 Kim JW, Lim SC, Lee MY, Lee JW, Oh WK, Kim SK, Kang KW: Inhibition of neointimal formation by trans-resveratrol: role of phosphatidyl inositol 3-kinase-dependent Nrf2 activation in heme oxygenase-1 induction. Mol Nutr Food Res 2010, Epub ahead of print.

25 Kode A, Rajendrasozhan S, Caito S, Yang SR, Megson IL, Rahman I: Resveratrol induces glutathione synthesis by activation of Nrf2 and protects against cigarette smoke-mediated oxidative stress in human lung epithelial cells. Am J Physiol Lung Cell Mol Physiol 2008;294:L478–L488.

26 Zhang H, Shih A, Rinna A, Forman HJ: Exacerbation of tobacco smoke mediated apoptosis by resveratrol: an unexpected consequence of its antioxidant action. Int J Biochem Cell Biol 2010, Epub ahead of print.

27 Rubiolo JA, Mithieux G, Vega FV: Resveratrol protects primary rat hepatocytes against oxidative stress damage: activation of the Nrf2 transcription factor and augmented activities of antioxidant enzymes. Eur J Pharmacol 2008;591:66–72.

28 Hsieh TC, Lu X, Wang Z, Wu JM: Induction of quinone reductase NQO1 by resveratrol in human K562 cells involves the antioxidant response element ARE and is accompanied by nuclear translocation of transcription factor Nrf2. Med Chem 2006; 2:275–285.

29 Bishayee A, Barnes KF, Bhatia D, Darvesh AS, Carroll RT: Resveratrol suppresses oxidative stress and inflammatory response in diethylnitrosamine-initiated rat hepatocarcinogenesis. Cancer Prev Res (Phila, Pa) 2010;3:753–763.

30 Saha P, Das S: Elimination of deleterious effects of free radicals in murine skin carcinogenesis by black tea infusion, theaflavins and epigallocatechin gallate. Asia Pac J Cancer Prev 2002;3:225–230.

31 Yuan JH, Li YQ, Yang XY: Inhibition of epigallocatechin gallate on orthotopic colon cancer by upregulating the Nrf2-UGT1A signal pathway in nude mice. Pharmacology 2007;80:269–278.

32 Zhang ZM, Yang XY, Yuan JH, Sun ZY, Li YQ: Modulation of NRF2 and UGT1A expression by epigallocatechin-3-gallate in colon cancer cells and BALB/c mice. Chin Med J 2009;122:1660–1665.

33 Andreadi CK, Howells LM, Atherfold PA, Manson MM: Involvement of Nrf2, p38, B-Raf, and nuclear factor-kappaB, but not phosphatidylinositol 3-kinase, in induction of hemeoxygenase-1 by dietary polyphenols. Mol Pharmacol 2006;69:1033–1040.

34 Na HK, Kim EH, Jung JH, Lee HH, Hyun JW, Surh YJ: (–)-Epigallocatechin gallate induces Nrf2-mediated antioxidant enzyme expression via activation of PI3K and ERK in human mammary epithelial cells. Arch Biochem Biophys 2008;476:171–177.

35 Sriram N, Kalayarasan S, Sudhandiran G: Epigallocatechin-3-gallate augments antioxidant activities and inhibits inflammation during bleomycin-induced experimental pulmonary fibrosis through Nrf2-Keap1 signaling. Pulm Pharmacol Ther 2009;22:221–236.

36 Romeo L, Intrieri M, D'Agata V, Mangano NG, Oriani G, Ontario ML, Scapagnini G: The major green tea polyphenol, (–)-epigallocatechin-3-gallate, induces heme oxygenase in rat neurons and acts as an effective neuroprotective agent against oxidative stress. J Am Coll Nutr 2009;28(suppl):492S–499S.

37 Shen G, Xu C, Hu R, Jain MR, Nair S, Lin W, Yang CS, Chan JY, Kong AN: Comparison of (–)-epigallocatechin-3-gallate elicited liver and small intestine gene expression profiles between C57BL/6J mice and C57BL/6J/Nrf2 (–/–) mice. Pharm Res 2005;22: 1805–1820.

38 Balogun E, Foresti R, Green CJ, Motterlini R: Changes in temperature modulate heme oxygenase-1 induction by curcumin in renal epithelial cells. Biochem Biophys Res Commun 2003;308:950–955.

39 Rushworth SA, Ogborne RM, Charalambos CA, O'Connell MA: Role of protein kinase C delta in curcumin-induced antioxidant response element-mediated gene expression in human monocytes. Biochem Biophys Res Commun 2006;341:1007–1016

40 McNally SJ, Harrison EM, Ross JA, Garden OJ, Wigmore SJ: Curcumin induces heme oxygenase 1 through generation of reactive oxygen species, p38 activation and phosphatase inhibition. Int J Mol Med 2007;19:165–172.

41 Nishinaka T, Ichijo Y, Ito M, Kimura M, Katsuyama M, Iwata K, Miura T, Terada T, Yabe-Nishimura C: Curcumin activates human glutathione S-transferase P1 expression through antioxidant response element. Toxicol Lett 2007;170:238–247.

42 Pae HO, Jeong GS, Jeong SO, Kim HS, Kim SA, Kim YC, Yoo SJ, Kim HD, Chung HT: Roles of heme oxygenase-1 in curcumin-induced growth inhibition in rat smooth muscle cells. Exp Mol Med 2007; 39:267–277.

43 Kang ES, Woo IS, Kim HJ, Eun SY, Paek KS, Kim HJ, Chang KC, Lee JH, Lee HT, Kim JH, Nishinaka T, Yabe-Nishimura C, Seo HG: Up-regulation of aldose reductase expression mediated by phosphatidylinositol 3-kinase/Akt and Nrf2 is involved in the protective effect of curcumin against oxidative damage. Free Radic Biol Med 2007;43:535–545.

44 Garg R, Gupta S, Maru GB: Dietary curcumin modulates transcriptional regulators of phase I and phase II enzymes in benzo[*a*]pyrene-treated mice: mechanism of its anti-initiating action. Carcinogenesis 2008;29:1022–1032.
45 Farombi EO, Shrotriya S, Na H-K, Kim SH, Surh Y-J: Curcumin attenuates dimethylnitrosamine-induced liver injury in rats through Nrf2-mediated induction of heme oxygenase-1. Food Chem Toxicol 2008;46:1279–1287.
46 Mandal MN, Patlolla JM, Zheng L, Agbaga MP, Tran JT, Wicker L, Kasus-Jacobi A, Elliott MH, Rao CV, Anderson RE: Curcumin protects retinal cells from light-and oxidant stress-induced cell death. Free Radic Biol Med 2009;46:672–679.
47 Suphim B, Prawan A, Kukongviriyapan U, Kongpetch S, Buranrat B, Kukongviriyapan V: Redox modulation and human bile duct cancer inhibition by curcumin. Food Chem Toxicol 2010; 48:2265–2272.

Prof. Young-Joon Surh
College of Pharmacy, Seoul National University
599 Kwanak-ro, Kwanak-gu
Seoul 151–742 (South Korea)
Tel. +82 2 880 7845, Fax +82 2 874–9775, E-Mail surh@plaza.snu.ac.kr

Naito Y, Suematsu M, Yoshikawa T (eds): Free Radical Biology in Digestive Diseases.
Front Gastrointest Res. Basel, Karger, 2011, vol 29, pp 97–110

Free Radicals and Gastric Mucosal Injury

Chan Young Ock · Kyung Sook Hong · Ju Hyun Kim · Ki-Baik Hahm

Department of Gastroenterology, Gachon Graduate School of Medicine Gil Medical Center and Laboratory of Translational Medicine, Gachon University of Science and Medicine Lee Gil Ya Cancer and Diabetes Institute, Incheon, South Korea

Abstract

Free radicals in excess high amounts, irrespective of radical and/or nonradical reactive species, are very reactive in nature and impose detrimental influence on living organisms through damaging all major cellular constituents including membrane, cytoplasmic proteins, and even nuclear DNAs, resulting in oxidative stress, whereas nitric oxide (NO), superoxide anion, and related reactive oxygen species (ROS) in low or modest amounts play an important role as regulatory mediators in signaling processes through which many of the ROS-mediated responses, paradoxically, can protect the cells against oxidative stress and maintain redox homeostasis. Compared to primitive organisms, vertebrates have evolved the use of NO and ROS also as signaling molecules for several physiological functions, which are generated by tightly regulated enzymes including NO synthase and NADPH oxidase isoforms, respectively. Simply stated, the problem is provoked by blazed ROS productions far beyond the quenching capacity of cells, by which free radicals are implicated in the pathogenesis of several gastrointestinal diseases, cancer, diabetes mellitus, atherosclerosis, neurodegenerative diseases, rheumatoid arthritis, ischemia and reperfusion injury, and other diverse diseases in addition to senescence. The implication of free radicals in gastric mucosal injuries will be the main focus of the current chapter. The offending systems like NADPH oxidase and NO synthase and defending systems of glutathione and heat shock protein related to gastric mucosal injury will be introduced.

The gastric mucosa is continuously challenged by a variety of aggressive factors of both endogenous and exogenous irritants, including excess secretion of gastric acids and pepsin, alcohol, reactive oxygen species (ROS), nonsteroidal anti-inflammatory drugs (NSAIDs), excess physical or mental stress, and *Helicobacter pylori* infection. To protect the gastric mucosa from these aggressive factors, a complex defense system has evolved including the production of surface mucus and bicarbonate, the regulation of gastric mucosal blood flow, the acceleration of epithelial regeneration, alkaline tide, afferent sensory neuron, and the preservation of epithelial homeostasis. Prostaglandin (PG), in particular PGE_2, bicarbonates, and heat shock protein (HSP) enhance these protective mechanisms and are therefore believed to comprise a major gastric mucosal defensive factor [1, 2]. Therefore, as much as PGs, HSPs and several proteins involved in scavenging ROS were proved to be another key protective mechanism. Since the imbalance between aggressive and defensive factors determines the outcomes of gastric lesions under the exposure to noxious etiologies represented with either a relative increase in aggressive factors or a considerable decrease in protective factors, the therapeutic intervention also consists of either attenuating offensive factors or enhancing defensive factors [3].

Therefore, ROS play a very fundamental and important role in the pathogenesis of gastroduodenal mucosal inflammation related to mucosal ischemic injury and other models of mucosal damage induced by NSAIDs, alcohol, or *H. pylori*. In detail, *H. pylori* achieve its pathogenetic role by triggering an intense leukocyte infiltration of the gastric mucosa and these neutrophil activations provide a major source of ROS. In the case of NSAID-associated gastric mucosal injury, adhesion molecules are increasingly expressed on the surface of endothelial cells, and leukocytes serve to ensure an orderly sequence of cell-to-cell interactions that sustain leukocyte adherence to vascular endothelium, leading to subsequent transendothelial migration into inflamed tissue. Similarly, alcohol-induced gastritis can predispose to gastric cancer development through overwhelming productions of ROS. Reflux of alkaline duodenal contents with cigarette smoking is an additional important injurious factor of endogenous origin, of which gastric mucosal injuries were also based on excessive ROS generations. In addition, mental or physical stress is among common exogenous mucosal irritants, though invisible, that can inflict mucosal injury [4, 5]. The ability of the stomach to defend itself against these noxious etiologies has been ascribed to a number of factors constituting the gastric mucosal defense, including mucus and bicarbonate secretions, PGs, sulfhydryl compounds, glutathione, HSP, and gastric mucosal blood flow. Summarized in brief, the balance between the ROS-generating and ROS-scavenging systems is one of the key points in gastric mucosal injury.

ROS and reactive nitrogen species (RNS) are well recognized for playing a dual role as both deleterious and beneficial species. ROS are oxygen-derived small molecules, including oxygen radicals (superoxide, hydroxyl, peroxyl, and alkoxyl) and certain nonradicals that are either oxidizing agents and/or are easily converted into radicals, such as hypochlorous acid (HOCl), ozone (O_3), singlet oxygen ($^1O_2$), and hydrogen peroxide (H_2O_2). Nitrogen-containing oxidants, such as nitric oxide (NO), are called RNS. ROS generation is generally a cascade of reactions that starts with the production of superoxide. ROS and RNS are normally generated by tightly regulated enzymes, such as NO synthase (NOS) and NADPH oxidase (NOX) isoforms, respectively. Overproduction of ROS results in oxidative stress, a deleterious process that can be an important mediator of damage to cell structures, including lipids and membranes, proteins, and DNA. On the other hand, beneficial effects of ROS/RNS can occur at low or moderate concentrations and involve physiological roles including defense against infectious agents, a number of cellular signaling pathways, and the induction of a mitogenic response. Ironically, various ROS-mediated actions in fact protect cells against ROS-induced oxidative stress and re-establish or maintain redox balance termed also redox homeostasis [6, 7]. Since the following chapters will deal with the description of free radicals' aspects according to etiology, including *H. pylori* and NSAIDs focusing on gastric mucosal injury, this chapter will describe: (1) gastric mucosal injury related to oxidative stress; (2) NOX and NOS intervention in generating ROS and RNS; (3) endogenous reduced glutathione (GSH) and HSP intervention in defense against oxidative stress; (4) the cross-link between ROS-associated gastric inflammation and carcinogenesis, and (5) the overview of antioxidant intervention as the treatment of gastric mucosal injury.

Gastric Irritants and Oxidative Stress of Their Pathogenic Mechanisms

Alcohol-Induced Oxidative Stress

The metabolism of alcohol is executed in both microsomal and mitochondrial systems of the cell, and directly produces ROS and RNS, attributable to oxidative stress as the basic injurious mechanism. In addition to ROS generation, alcohol results in the depletion of GSH levels, thus attenuating endogenous antioxidant activity [8, 9]. Moreover, alcohol elevates malondialdehyde, hydroxyethyl radical, and hydroxynonenal protein adducts, leading to the modification of all biological structures and consequently resulting in serious malfunction of cells and tissues, finally leading to organ dysfunction and even cancer development. Acute and chronic alcohol treatment increases the production of ROS, lowers cellular antioxidant levels, and enhances oxidative stress in many tissues, especially in the liver and stomach. Therefore, alcohol-induced oxidative stress plays a major role in the mechanisms by which ethanol induces liver injury including alcoholic steatohepatitis, cirrhosis, and hepatocellular carcinoma and provokes gastritis and ulcer. Alcohol also facilitates the development of gastroesophageal reflux disease by reducing the pressure of the lower esophageal sphincter and weakening esophageal motility, and definitely alcohol facilitates the development of oropharyngeal, esophageal, gastric, and colon cancer. Simply stated, drinking alcohol is the beginning step disturbing GI tract homeostasis, so called redox imbalance.

Smoking-Induced Oxidative Stress

Smoking and nicotine have been acknowledged to have significant adverse effects on peptic ulcer disease. Smoking and chronic nicotine treatment stimulate basal acid output mediated through the stimulation of histamine type 2 receptor by histamine released after mast cell degranulation and pepsinogen secretion by increasing chief cell number or with an enhancement of their secretory capacity. Smoking also increases bile salt reflux rate and gastric bile salt concentration, thereby increasing duodenogastric reflux that augments the risk of gastric ulcer in smokers in addition to duodenal ulcer-related increased acid output. Mechanistically, smoking increases ulcerogenesis through oxidative damage of the mucosa by increasing the generation of ROS potentiated by nicotine and smoking. Smoking increases production of platelet-activating factor and endothelin-1, which are potent gastric ulcerogens via vasoconstriction. Cigarette smoking and nicotine reduce the level of circulating epidermal growth factor (EGF) and decrease the secretion of EGF from the salivary gland, which are necessary for gastric mucosal cell renewal. Nicotine also decreases PG generation in the gastric mucosa of smokers, thereby making the mucosa susceptible to ulceration. Among these pathogeneses, ROS generation and ROS-mediated gastric mucosal cell apoptosis are also considered to be important mechanisms for aggravation of ulcer by cigarette smoke or nicotine. Furthermore, both smoking and nicotine reduce angiogenesis in the gastric mucosa through inhibition of NO synthesis, thereby arresting cell renewal process. Simply stated, smoking makes the oxidative defense system futility.

Stress-Induced Oxidative Stress

Stress-related mucosal disease (SRMD) and subsequent upper GI bleeding remain significant concerns in critically ill patients. Although several factors, for instance, hypoperfusion of the GI tract, reflux of bile salts, inflammation prone status, and increased vulnerability to damaging causes, etc, contribute to stress-related ulceration, acid is presumed to be a major contributor to SRMD, by which fundamental treatment strategies use H2-RAs and proton pump inhibitors (PPIs). However, oxidative stress has been speculated as a major contributing factor to SRMD based on the fact that acid suppressants only contributed to rescuing patients from catastrophe, but not be completely free from SRMD, as evidenced by high mortality, high recurrence, and deranged organ dysfunction in addition to unexpected therapeutic efficacy with antioxidants. Though exact pathophysiology of SRMD is not known due to multifactorial causes and a complex set of interactions that cause a breakdown of mucosal proactive defenses leading to ulceration, ROS is known to be principally involved in SRMD. Despite evidence that *H. pylori* infection is closely associated with stress in gastric ulcer patients, the underlying mechanism why ulcer recurrence after stress is augmented especially in patients with *H. pylori* remains unknown. In this condition, we hypothesized that oxidative stress played a critical role in the augmented mucosal damage provoked by water immersion restraint stress (WIRS) in *H. pylori* infection and that the antioxidant α-tocopherol could ameliorate the aggravation of stress-associated gastric mucosal damage [10]. Remarkably increased hemorrhagic lesions and bleeding indexes were noted in the *H. pylori*-infected group with statistical significance ($p < 0.05$) compared to the noninfected group with the same duration of WIRS. Significantly higher oxidative stress documented by inducible NOS (iNOS), lipid peroxides, and GSH level was detected in gastric homogenates of the *H. pylori*-infected group. The fact that α-tocopherol pretreatment significantly prevented the gastric mucosal damage caused by WIRS in the presence of *H. pylori* suggested that the presence of *H. pylori* caused significant deterioration of stress-induced gastric mucosal lesions through increased oxidative stress, and thus antioxidant treatment such as α-tocopherol protected the gastric injuries.

H. pylori, NSAIDs, and Other Causes of Oxidative Stress

Infection with *H. pylori* plays a role in the pathogenesis of gastritis, peptic ulcer, gastric carcinoma, and gastric lymphoma, but mechanisms leading to the various clinical manifestations still remain obscure. Presumptively, excessive apoptosis or proliferation inhibition will result in cell mass loss, which is frequently observed in chronic atrophic gastritis (CAG) and ulcers. There has been enough scientific evidence that demonstrate an association between *H. pylori* infection and exacerbated synthesis of free radicals, the latter being also well known as a primary cause of cell death underlying *H. pylori*-associated CAG, ulcer, and cancer. *H. pylori* are able to induce polymorphonuclear and mononuclear cells producing ROS that could cause DNA damage to the adjacent cells leading to cancer development [11]. Thus, *H. pylori*-induced oxidative stress activates inflammatory bouts in addition to the intrinsic pathway of apoptosis. NSAIDs such as aspirin and indomethacin are known to induce gastric mucosal damage like bleeding, ulceration, and perforation in humans and animals. In addition to a deficiency of endogenous PGs due to inhibition of cyclooxygenase by the drug, excess ROS generation may be an important prime event that leads to mucosal injury induced by NSAIDs. For instance, lipid peroxidation (LPO) mediated by oxygen radicals, especially hydroxyl radicals, plays a crucial role in the development of the gastric mucosal injury induced by indomethacin.

Lipid Peroxidation, NADPH Oxidase and NO Synthase

Lipid Peroxidation and Oxidative DNA Damage

Chronic inflammation induced by biological, chemical, and physical factors has been associated with increased risk of human cancer at various sites. This is through oxidative/nitrosative stress and LPO, thereby generating excess ROS, RNS, and DNA-reactive aldehydes. Miscoding etheno- and propano-modified DNA bases are generated by reaction of DNA with these major LPO products. Persistent oxidative stress and excess LPOs induced by inflammatory processes, impaired metal storage, and/or dietary imbalance cause accumulations and massive DNA damage. This massive DNA damage, along with deregulation of cell homeostasis, leads to carcinogenesis. Therefore, steady-state levels of DNA damage caused by ROS, RNS, and LPO end products provide promising molecular signatures for risk prediction. We have measured the changes of 8-hydroxydeoxyguanosine (8-OHdG) contents of DNA from human gastric mucosa with or without *H. pylori* accompanied with the measurements of two additional biomarkers, iNOS and apoptosis, in gastric biopsies obtained before and after the eradication of *H. pylori*. The 8-OHdG contents of patients with positive *H. pylori* were significantly increased compared with those of normal controls with negative *H. pylori* ($p < 0.01$). Moreover, after the eradication of *H. pylori*, both the apoptotic index and the iNOS scores were significantly decreased compared with those before eradication in parallel with a decrement in 8-OHdG levels. The increased levels of oxidative DNA damage, increased occurrence of apoptosis, and increased expression of iNOS suggest mechanistic links between *H. pylori* infection and gastric carcinogenesis [12].

NAPDH Oxidases

NOX family serves a variety of functions including ROS generation, antimicrobial defense, biosynthetic processes, oxygen sensing, and cellular signaling. These enzymes share the capacity to transport electrons across the plasma membrane and to generate superoxide and other downstream ROS (fig. 1). NOX1 and NOX3 are similar to the NOX2-based phagocytic oxidase, for which cytosolic activator and organizer proteins like NOXA, NOXO, and GTP-Rac are required. Determinants of subcellular targeting include formation of NOX-p22^{phox} heterodimeric complexes allowing plasma membrane translocation, phospholipid-binding specificities of PX domain-containing organizer proteins [p47^{phox} or NOX organizer 1 (NOXO1) and p40^{phox}], and variably splicing of Noxo1 PX domains. Dual oxidases (DUOX1 and DUOX2) are targeted by different mechanisms. For a rather long time, superoxide generation by NOX was considered as an oddity only found in professional phagocytes. Later on, in addition to innate host defense, diverse physiologic actions including posttranslational processing of proteins, cellular signaling, gene regulation, and cell differentiation had been explored in many other cell types like fibroblasts, vascular smooth muscle cells, and even GI epithelial cells [13]. Especially in the stomach, NOX enzymes also contribute to a wide range of pathological processes like the initiation, propagation and exacerbation of gastric inflammation, and consequent ulceration. As a sign of association with gastric mucosal injury, *H. pylori* infection has been suggested to stimulate expression of the NOX1-based oxidase system in guinea pig gastric epithelium, and NOX1-base oxidase may be a potential marker of neoplastic transformation and play an important role in oxygen radical- and inflammation-dependent carcinogenesis in the human stomach [14].

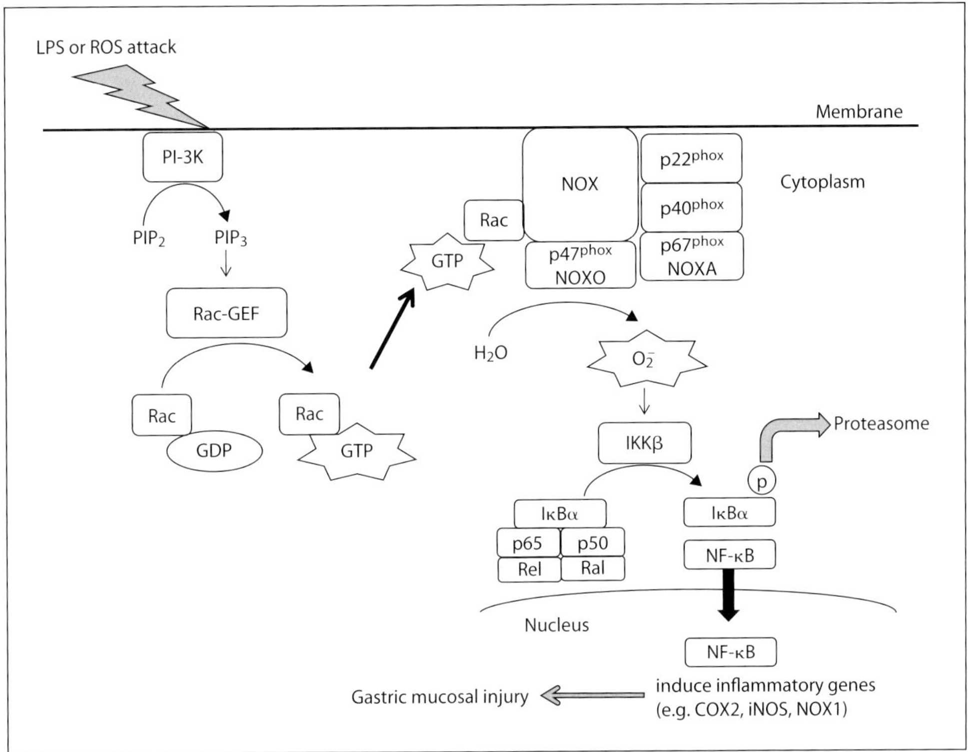

Fig. 1. NOX activation and subsequent ROS generation in stress-induced gastric mucosal damage. The sequential activations of PI-3K, GEF, Rac-GEF led to NOX activation, after which superoxide generation initiated propagated chain reaction for more reactive ROS. These ROS activations were responsible for activating redox-sensitive transcription factor NF-κB, explaining the association between free radicals and gastric mucosal injuries after insults.

Nitric Oxide Synthase

Accumulating evidence indicates that free radical species and NO or its derivatives are the key denominators in carcinogenesis. For instance, under the hypothesis that overproduction of NO via iNOS is suggested to be a significant pathogenic factor in *H. pylori*-induced gastritis, Nam et al. [15] studied the role of iNOS in *H. pylori*-associated gastric carcinogenesis, for which gastric cancer was generated in mice using a combination treatment comprising N-methyl-N-nitrosourea administration and *H. pylori* infection in iNOS-deficient mice and wild-type littermates. The overall incidence of gastric cancer at week 50 was significantly lower in $iNOS^{-/-}$ compared with wild-type littermates ($p < 0.05$), suggesting that iNOS contributes to *H. pylori*-associated gastric carcinogenesis in mice. The three enzymatic sources of NO, neuronal NO synthase (nNOS), endothelial NOS (eNOS), and iNOS, have been characterized in the gastrointestinal tract. NO derived from constitutive NO synthases (eNOS and nNOS) has been regarded as having a protective role, whereas iNOS has been paradoxically implicated in detrimental roles like inflammation and carcinogenesis in the stomach. This is due to the accumulated evidence of its interaction with reactive species producing peroxynitrites ($ONOO^-$) and other compounds, and it was found to be overexpressed in chronic inflammatory diseases and various types of cancer. However, since NO plays a multi-faceted role in gastric mucosal integrity in addition to being an endogenous modulator of numerous physiological functions like the regulation of blood flow, maintenance of vascular tone, control of platelet aggregation, central and peripheral neurotransmission, and mastocyte activity, the ultimate effects of NO should be interpreted with caution [16]. In summary, NO is a crucial mediator of gastrointestinal mucosal defense, but, paradoxically, it can contribute to gastric mucosal injury in parallel to increased iNOS activities.

Gastric Protectants Focused on Attenuating Oxidative Stress: Glutathione, Heat Shock Protein and *NF-E2-Related Factor 2*

Endogenous Reduced Glutathione

The primary role of reduced glutathione (L-γ-glutamyl-L-cysteinyl-glycine, GSH), the prevalent low-molecular-weight thiol in mammalian cells, is to protect cells from oxidative stress. It is abundantly distributed in the GI tract mucosal cells, and its highest concentration is found in the duodenum. Glutathione tripeptide is central to a complex multifaceted detoxification system, where there is substantial interdependence between separate component members. Glutathione tripeptide is formed in a two-step enzymatic process; the first step involves the formation of γ-glutamylcysteine from glutamate and cysteine by the activity of the γ-glutamylcysteine synthetase, and the second step is the formation of GSH by the activity of GSH synthetase which uses γ-glutamylcysteine and glycine as substrates. Glutathione participates in detoxification at several different levels, and may scavenge free radicals, reduce peroxides or be conjugated with electrophilic compounds. Thus, glutathione provides the cell with multiple defenses not only against ROS, but also against the other kinds of toxic products. The detoxifying capability of glutathione is directly related to its thiol group and to its function as a substrate for enzymatic activity; in fact, glutathione regulates the action of glutathione peroxidases and glutathione transferases. GI glutathione peroxidase (GPX) is the fourth member of the GPX family. Alterations in its concentration have also been demonstrated to be a common feature of many pathological conditions including diabetes, cancer, AIDS, neurodegenerative, and liver diseases, among which gastric mucosal damage after *H. pylori* infection is intimately associated. As shown in figure 2, the inflammatory activation after *H. pylori* infection is significantly correlated with GSH amount,

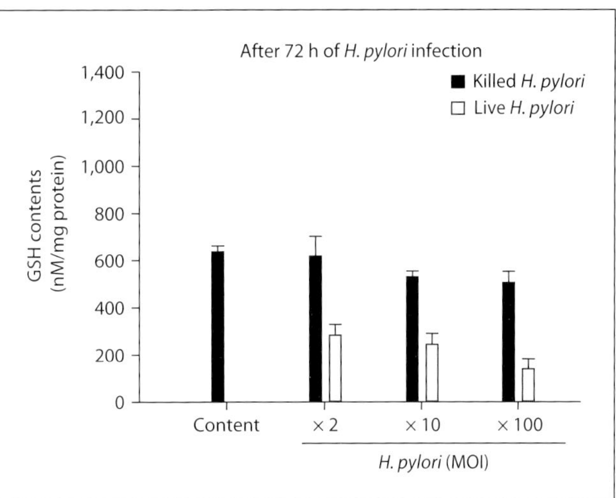

Fig. 2. Association between *H. pylori* infection and host cell glutathione contents. When the cells were infected with *H. pylori*, they showed gradual depletion of GSH in a time-dependent manner, suggesting that the cancellation of GSH leads to cells vulnerable to oxidative stress with *H. pylori* infection. Interestingly, only live *H. pylori* were responsible for depleted GSH levels.

suggesting the depletion of GSH might be one of pathogenic mechanism after *H. pylori* infection [17]. Therefore, the therapeutic regulation of glutathione availability could provide a novel method for preventing or reducing the damage caused during *H. pylori* infection because there is a significant correlation between the degrees of *H. pylori*-associated gastric inflammation and GSH concentration. Key transcription factors identified thus far include *NF-E2-related factor 2 (Nrf-2)/Nrf-1* via the antioxidant response element, activator protein-1 and NF-κB. In summary, GSH is the most abundant intracellular antioxidant thiol and is central to redox defense during oxidative stress.

Heat Shock Proteins

The first level of gastric mucosal barrier consists of the factors secreted into the lumen including bicarbonates, mucus, immunoglobulins, other antibacterial substances including lactoferrin, and surface active phospholipids. The second level of the defense system is the gastric epithelia, which are remarkably resistant to acids or irritants and form a relatively tight barrier to passive diffusion. In addition, the epithelium is capable of undergoing extremely rapid repair and restitution if its continuity is disrupted. The third level of gastric mucosal barrier is the mucosal microcirculation in concert with sensory afferent nerves within the mucosa and submucosa. Back diffusion of acid or toxin into the mucosa results in neural system-mediated elevations of calcitonin gene-related peptide, which contribute to enhancing mucosal blood flow that is critical for limiting damage and facilitating repair. The fourth level of defense is the mucosal immune system consisting of mast cells and macrophage, which orchestrate an appropriate inflammatory response to challenge. All the above factors are known to contribute to the orchestrated artwork of 'gastric mucosal protection'. In recent years, HSPs have been revealed to be an additional factor utilized for the gastric defense mechanisms at the intracellular level. Certain HSPs are expressed under non-stressful conditions and play an important role in the maintenance of normal cell integrity, but certain HSPs improve cellular recovery by either refolding partially damaged functional proteins or increasing delivery of precursor proteins to important organelles such as mitochondria and endoplasmic reticulum through which HSPs might complete efficient mucosal defense mechanisms and achieve efficient ulcer healing, most probably protecting key enzymes related to cytoprotection [18].

NF-E2-Related Factor 2 (Nrf-2)

Electrophile-responsive element-mediated gene induction is a pivotal mechanism of cellular defense against the toxicity of electrophiles and ROS. *Nrf-2* was demonstrated to regulate the induction of genes encoding antioxidant proteins and phase 2 detoxifying enzymes. Under normal conditions, *Nrf-2* localizes in the cytoplasm where it interacts with the actin-binding protein, Kelch-like ECH associating protein 1 (Keap1), and is rapidly degraded by the ubiquitin-proteasome pathway. Signals from ROS or electrophilic insults target the *Nrf-2*-Keap1 complex, dissociating *Nrf-2* from Keap1. Stabilized *Nrf-2* then translocates to the nuclei and transactivates its target genes [19]. Constitutive activation of *Nrf-2*-regulated transcription in Keap1$^{-/-}$ mice clearly demonstrated that the disruption of Keap1 repression is sufficient for the activation of *Nrf-2*. Briefly summarized, the mechanism that modulates *Nrf-2*-Keap1 interaction is pivotal for the cellular sensing mechanism for electrophiles.

Oxidative Stress, Gastric Mucosal Injury and Progress to Carcinogenesis

Growing evidence suggests that ROS within cells act as second messengers in intracellular signaling cascades which can induce the oncogenic phenotype of cancer cells by virtue of their ability to increase cell proliferation, survival, cellular migration, and also by inducing DNA damage leading to genetic lesions. In a similar fashion, chronic persisting inflammation contributes to cancer development. Taken together, the coworks of oxidative stress and gastric inflammation might provide the basis to carcinogenesis [20]. This bad combination is exemplified in *H. pylori*-associated gastritis, refluxate-induced gastroesophageal reflux disease, and chronic colitis, all associated with gastric cancer, esophageal cancer, and colitic cancer, supporting the fact that ROS/RNS geared the journey to carcinogenesis in affected organs. Hallmarks of cancer-associated inflammation include the presence of infiltrating leukocytes,

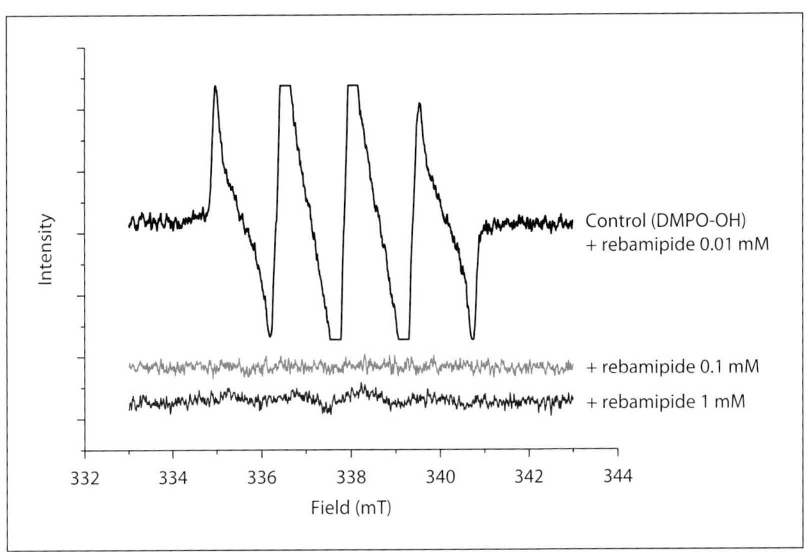

Fig. 3. Scavenging action of rebamipide (Mucosta®) documented with electron spin resonance. DMPO-adducted spin of hydroxyl radicals is shown. Rebamipide >0.1 mM showed significant scavenging action against hydroxyl radicals. Rebamipide (0.01–1 mM) was mixed in 200 µl containing 0.05 mM $FeSO_4$, 1 mM H_2O_2, 1 mM 5,5-dimethylpyrroline-N-oxide (DMPO) and 50 mM sodium phosphate, pH 7.4, at 37°C. Reactions were initiated by adding H_2O_2. After 1 min, aliquots of the reactions were transferred to crystal cuvettes. The spectrum of DMPO-OH was displayed using ESR spectrophotometer.

cytokines, chemokines, growth factors, lipid messengers, and matrix-degrading enzymes [21].

Antioxidants and Some Drugs Prescribed in Gastroenterology in Order to Rescue Oxidative Stress-Associated Gastric Mucosal Injury

Rebamipide

Rebamipide, a gastroprotective drug, was developed in Japan for the treatment of peptic ulcer disease. Rebamipide was proven superior to the former same category drug cetraxate in a randomized, controlled, double-blind, comparative clinical study. The drug's mechanisms of action are different from antisecretory drugs; it accelerates and improves the quality of ulcer healing and reduces ulcer recurrence rate. This is mainly thanks to its PG induction action and ROS scavenging activity (fig. 3). A number of basic and translational researches have been performed to clarify the mechanisms of rebamipide's action [22, 23]. These studies demonstrated unique properties of rebamipide and convincingly showed that it increases gastric mucus glycoprotein components, stimulates migration and proliferation of wounded epithelial cell monolayers, increases expression of EGF and its receptor in normal and ulcerated gastric mucosa, and scavenges active oxygen radicals. Based on these pharmacological actions, it protects the gastric mucosa against acute injury caused by various noxious and ulcerogenic factors. Inhibition of immunoinflammatory responses by rebamipide in *H. pylori*-infected patients may prevent development of gastritis, peptic ulcer disease, its recurrence, and possibly gastric cancer.

Proton Pump Inhibitors

Although PPIs are potent blockers of gastric acid secretion and are widely regarded as drug of choice for the treatment of acid-peptic disorders, PPIs have been found to have additional antioxidant properties and direct anti-inflammatory effects on neutrophils, monocytes, endothelial, and epithelial cells. Those anti-inflammatory effects of the PPIs might influence a variety of inflammatory disorders, both peptic and non-peptic, within and outside of the gastrointestinal tract [24]. Recently, we published data showing that significant anti-inflammatory, antioxidative, and antimutagenic activities of PPI played a cancer-preventive role against colitis-induced carcinogenesis. This novel in vivo evidence is suggestive of chemopreventive action independent of gastric acid suppression.

Phytoceuticals

Phytoceutical is a term for plant products that are active on biological systems. Phytoceuticals such as Korea red ginseng, green tea, red wine, flavonoids, broccoli sprouts, garlic, and probiotics are known to inhibit *H. pylori* colonization, decrease gastric inflammation by inhibiting cytokine and chemokine release, and repress precancerous changes by inhibiting NF-κB DNA binding, inducing profuse levels of apoptosis, and inhibiting mutagenesis [25]. Even though further unsolved issues are still awaited before phytoceuticals are accepted as a standard treatment for *H. pylori* infection, phytoceuticals can be a mighty weapon for either suppressing or modulating the disease-associated footprints of *H. pylori* infection. Korea red ginseng is a good example of a natural herb with ubiquitous properties conductive to stopping inflammatory carcinogenesis that is associated with augmented *H. pylori* eradication, reversal of CAG and intestinal metaplasia, and maintenance of homeostasis based on its antioxidative actions [26]. Flavonoids, a highly diverse class of secondary metabolites with potentially beneficial human health effects that is widely distributed in the plant kingdom and currently consumed in large amounts in the diet, display several pharmacological properties in the gastroprotective area, acting as antisecretory, cytoprotective and antioxidant agents.

Special Extracts of Licorice

Licorice, one of well-known traditional medicines in East Asia, has a natural sweet taste, various pharmacologic applications in detoxification, anti-inflammation, anti-cancer, anticoagulation, and an antimicrobial effect. Glycyrrhizin, isoliquiritigenin and licochalcone A are known as the main effector component of licorice [27]. Special extracts of licorice called 'Ad-lico', showed potent anti-inflammatory and anticancer efficacy in in vitro *H. pylori*-induced gastric inflammation and in vivo gastric cancer model according to concentration, potent antioxidative and anti-inflammatory effects in lower doses of <20 μg/ml, and antiproliferative and anticancer effects in high doses of >50 μg/ml, raising the hope that these novel extracts of licorice can be the ultimate drug for the treatment of *H. pylori*-associated gastritis and gastric cancer.

Acknowledgement

This study was supported by grants from the Ministry for Education, Science and Technology, Republic of Korea.

References

1 Robert A, Nezamis JE, Lancaster C, Hanchar AJ: Cytoprotection by prostaglandins in rats. Prevention of gastric necrosis produced by alcohol, HCl, NaOH, hypertonic NaCl, and thermal injury. Gastroenterology 1979;77:433–443.

2 Laine L, Takeuchi K, Tarnawski A: Gastric mucosal defense and cytoprotection: bench to bedside. Gastroenterology 2008;135:41–60.

3 Brzozowski T, Konturek PC, Konturek SJ, Brzozowska I, Pawlik T: Role of prostaglandins in gastroprotection and gastric adaptation. J Physiol Pharmacol 2005;56(suppl 5):33–55.

4 Robert A, Nezamis JE, Lancaster C, Davis JP, Field SO, Hanchar AJ: Mild irritants prevent gastric necrosis through 'adaptive cytoprotection' mediated by prostaglandins. Am J Physiol 1983;245:G113–G121.

5 Tsukimi Y, Okabe S: Recent advances in gastrointestinal pathophysiology: role of heat shock proteins in mucosal defense and ulcer healing. Biol Pharm Bull 2001;24:1–9.

6 Hussain SP, Hofseth LJ, Harris CC: Radical causes of cancer. Nat Rev Cancer 2003;3:276–285.

7 Federico A, Morgillo F, Tuccillo C, Ciardiello F, Loguercio C: Chronic inflammation and oxidative stress in human carcinogenesis. Int J Cancer 2007;121:2381–2386.

8 Lee JS, Oh TY, Kim YK, Baik JH, So S, Hahm KB, Surh YJ: Protective effects of green tea polyphenol extracts against ethanol-induced gastric mucosal damages in rats: stress-responsive transcription factors and MAP kinases as potential targets. Mutat Res 2005;579:214–224.

9 Hiraishi H, Shimada T, Ivey KJ, Terano A: Role of antioxidant defenses against ethanol-induced damage in cultured rat gastric epithelial cells. J Pharmacol Exp Ther 1999;289:103–109.

10 Oh TY, Yeo M, Han SU, Cho YK, Kim YB, Chung MH, Kim YS, Cho SW, Hahm KB: Synergism of *Helicobacter pylori* infection and stress on the augmentation of gastric mucosal damage and its prevention with alpha-tocopherol. Free Radic Biol Med 2005;38:1447–1457.

11 Park S, Kim WS, Choi UJ, Han SU, Kim YS, Kim YB, Chung MH, Nam KT, Kim DY, Cho SW, Hahm KB: Amelioration of oxidative stress with ensuing inflammation contributes to chemoprevention of H. pylori-associated gastric carcinogenesis. Antioxid Redox Signal 2004;6:549–560.

12 Hahm KB, Lee KJ, Choi SY, Kim JH, Cho SW, Yim H, Park SJ, Chung MH: Possibility of chemoprevention by the eradication of *Helicobacter pylori*: oxidative DNA damage and apoptosis in H. pylori infection. Am J Gastroenterol 1997;92:1853–1857.

13 Bokoch GM, Knaus UG: NADPH oxidases: not just for leukocytes anymore! Trends Biochem Sci 2003; 28:502–508.
14 Nakagiri A, Murakami M: Roles of NADPH oxidase in occurrence of gastric damage and expression of cyclooxygenase-2 during ischemia/reperfusion in rat stomachs. J Pharmacol Sci 2009;111:352–360.
15 Nam KT, Hahm KB, Oh SY, Yeo M, Han SU, Ahn B, Kim YB, Kang JS, Jang DD, Yang KH, Kim DY: The selective cyclooxygenase-2 inhibitor nimesulide prevents *Helicobacter pylori*-associated gastric cancer development in a mouse model. Clin Cancer Res 2004;10:8105–8113.
16 Jaiswal M, LaRusso NF, Gores GJ: Nitric oxide in gastrointestinal epithelial cell carcinogenesis: linking inflammation to oncogenesis. Am J Physiol Gastrointest Liver Physiol 2001;281:G626–G634.
17 Matthews GM, Butler RN: Cellular mucosal defense during *Helicobacter pylori* infection: a review of the role of glutathione and the oxidative pentose pathway. Helicobacter 2005;10:298–306.
18 Choi SR, Lee SA, Kim YJ, Ok CY, Lee HJ, Hahm KB: Role of heat shock proteins in gastric inflammation and ulcer healing. J Physiol Pharmacol 2009;60(suppl 7):5–17.
19 Itoh K, Tong KI, Yamamoto M: Molecular mechanism activating Nrf2-Keap1 pathway in regulation of adaptive response to electrophiles. Free Radic Biol Med 2004;36:1208–1213.
20 Lin WW, Karin M: A cytokine-mediated link between innate immunity, inflammation, and cancer. J Clin Invest 2007;117:1175–1183.
21 Hussain SP, Harris CC: Inflammation and cancer: an ancient link with novel potentials. Int J Cancer 2007;121:2373–2380.
22 Hahm KB, Park IS, Kim YS, Kim JH, Cho SW, Lee SI, Youn JK: Role of rebamipide on induction of heat-shock proteins and protection against reactive oxygen metabolite-mediated cell damage in cultured gastric mucosal cells. Free Radic Biol Med 1997;22:711–716.
23 Arakawa T, Kobayashi K, Yoshikawa T, Tarnawski A: Rebamipide: overview of its mechanisms of action and efficacy in mucosal protection and ulcer healing. Dig Dis Sci 1998;43:5S-13S.
24 Biswas K, Bandyopadhyay U, Chattopadhyay I, Varadaraj A, Ali E, Banerjee RK: A novel antioxidant and antiapoptotic role of omeprazole to block gastric ulcer through scavenging of hydroxyl radical. J Biol Chem 2003;278:10993–11001.
25 Lee SY, Shin YW, Hahm KB: Phytoceuticals: mighty but ignored weapons against *Helicobacter pylori* infection. J Dig Dis 2008;9:129–139.
26 Park S, Yeo M, Jin JH, Lee KM, Kim SS, Choi SY, Hahm KB: Inhibitory activities and attenuated expressions of 5-LOX with red ginseng in *Helicobacter pylori*-infected gastric epithelial cells. Dig Dis Sci 2007;52:973–982.
27 Oh HM, Lee S, Park YN, Choi EJ, Choi JY, Kim JA, Kweon JH, Han WC, Choi SC, Han JK, Son JK, Lee SH, Jun CD: Ammonium glycyrrhizinate protects gastric epithelial cells from hydrogen peroxide-induced cell death. Exp Biol Med (Maywood) 2009; 234:263–277.

Prof. Ki-Baik Hahm
Laboratory of Translational Medicine, Lee Gil Ya Cancer and Diabetes Institute, Gachon University of Medicine and Science
7–45 Songdo-dong, Yeonsu-gu
Incheon 406–840 (South Korea)
Tel. +82 32 899 6055, Fax +82 32 899 6054, E-Mail hahmkb@hotmail.com

Naito Y, Suematsu M, Yoshikawa T (eds): Free Radical Biology in Digestive Diseases.
Front Gastrointest Res. Basel, Karger, 2011, vol 29, pp 111–120

Free Radicals in *Helicobacter pylori* Infection

Hidekazu Suzuki[a] · Toshihiro Nishizawa[b] · Hitoshi Tsugawa[a] · Toshifumi Hibi[a]

[a]Division of Gastroenterology and Hepatology, Department of Internal Medicine, Keio University School of Medicine, [b]Department of Gastroenterology, National Tokyo Medical Center, Tokyo, Japan

Abstract

Helicobacter pylori infection, which is the main cause of gastritis, peptic ulcer disease and gastric cancer, is associated with infiltration of the gastric mucosa by neutrophils, macrophages, and B and T lymphocytes. However, this immune and inflammatory response cannot completely control the bacterial infection, and leaves the host prone to complications resulting from persistent inflammation. As a result, *H. pylori* infection may cause chronic inflammation, accumulation of reactive oxygen species, and oxidative DNA damage in the gastric mucosa. The *H. pylori* bacterial effector proteins are transported into the gastric host cells *via* the type IV secretory system, and regulate intracellular signal transduction. This mechanism provides novel insight into how *H. pylori* survives in the acidic environment of the human stomach. In cases with persistent gastric infection, the chronic gastritis may remain asymptomatic or may evolve into more severe diseases, such as peptic ulcer disease and chronic atrophic gastritis. In addition, infection with *H. pylori* increases the risk of development of gastric cancer and mucosa-associated lymphoid tissue lymphoma. This review focuses on the oxidative mechanisms involved in the bacterial and host-mucosal responses during colonization of the gastric mucosa by *H. pylori*.

Helicobacter pylori infection is now recognized as a major cause of not only gastric diseases such as chronic gastritis, peptic ulcer disease and gastric cancer, but also of nongastric diseases such as idiopathic thrombocytopenic purpura and chronic urticaria [1, 2]. Last year, the Japanese Society for *Helicobacter* Research announced that based on all the accumulated scientific evidence, *H. pylori* eradication should be undertaken in all infectants. In such *H. pylori*-associated diseases, formation of reactive oxygen species (ROS) and regulation of the oxidative stress in the bacteria as well as the host-gastric mucosa would be important pathological mechanisms.

In the present review, the oxidative stress on the bacterial and host-mucosal sides have been sequentially described and key issues discussed.

Bacterial Side Oxidative Stress

H. pylori strains have been classified based on their expression of the *cagA* gene and *cag* PAI, as well as on the *vacA* genotype. Strains with *cagA* have been shown to be associated with more severe gastric inflammation, increased bacterial load, and both peptic ulcer and gastric cancer. The *cag* PAI status affects the oxidative DNA damage and gastric epithelial apoptosis [3]. Ding et al. [4] indicated that the *cag* PAI status influenced the ability of *H. pylori* to induce intracellular ROS formation in the gastric epithelial cells.

Although persistent colonization of the gastric mucosa depends on the ability of the bacteria to respond to changing environmental conditions and to coordinate the expression of virulence factors during the course of infection, *H. pylori* possesses relatively few transcriptional regulators. According to the report by Barnard et al. [5], CsrA is necessary for full-range motility and survival of *H. pylori* under conditions of oxidative stress. Loss of *csrA* expression was reported to induce dysregulation of the oxidant-induced transcriptional responses of *napA* and *ahpC* (encoding alkyl hydroperoxide reductase), acid induction of *napA*, *cagA*, *vacA*, the urease operon and *fur*, as well as the heat shock response of *napA*, *groESL* and *hspR*. The authors concluded that CsrA has a broad role in regulating the physiology of *H. pylori* in response to environmental stimuli, and may be important in facilitating adaptation of the bacterium to the different environments encountered during colonization of the gastric mucosa.

Antioxidant Systems of *H. pylori*

During the process of colonizing the host, *H. pylori* induces a strong inflammatory response triggered by the infiltration of host cells. This defense is mediated by neutrophils and macrophages, culminating in the generation of large amounts of ROS and reactive nitrogen species that the persistent pathogens are exposed to [6–9]. Despite the oxidative stress induced in *H. pylori*, the bacterium survives and colonizes the gastric mucosa. Therefore, mechanisms for detoxification of ROS and repair of damaged cell components in *H. pylori* are of particular interest in understanding the pathogenicity of *H. pylori* and its capacity to induce persistent infection. It is well known that the *H. pylori* gene encodes superoxide dismutase (SOD) and catalase (KatA) as antioxidants and that its antioxidant ability is mainly mediated by these detoxification enzymes. Herein, the antioxidant system of *H. pylori* is discussed, mainly focusing on the role of these detoxification enzymes.

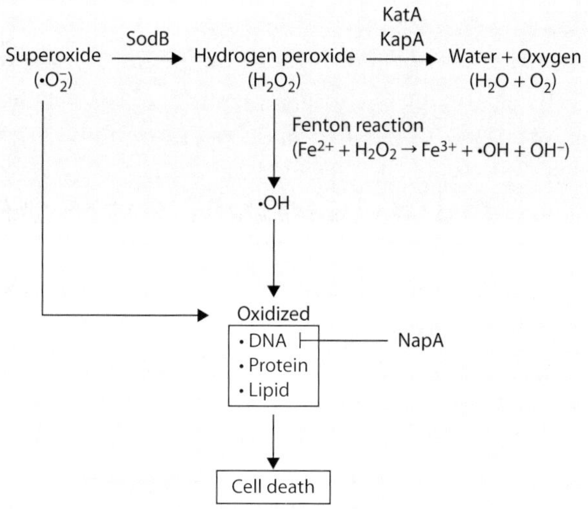

Fig. 1. The antioxidant system of *H. pylori*. Iron-cofactored SOD of *H. pylori* (SodB) catalyzes the dismutation of superoxide to oxygen and hydrogen peroxide, and then KatA and KapA catalyze the decomposition of H_2O_2 into H_2O and O_2 to protect cells from the oxidative damage. Furthermore, NapA binds to the DNA to protect the bacterium from oxidative damage.

Superoxide Dismutase of *H. pylori*

Superoxides (O_2^-) are not only produced by the strong immune responses, but are also generated during bacterial respiration and metabolism [10]. SOD catalyzes the dismutation of superoxide to oxygen and hydrogen peroxide in a cyclic redox reaction, and the hydrogen peroxide formed is subsequently removed by KatA or peroxidase. SODs are metallo-enzymes and three structurally different forms have been identified, depending on the metal cofactor. The Cu/Zn-SODs are generally located in the cytoplasm of eukaryotic cells and chloroplasts of some plants. Ni-SODs have been identified in several bacteria [11, 12]. Fe/Mn-SODs are prevalent in prokaryotes and in eukaryotic mitochondria. Organisms can carry a set of different SOD enzymes. *Escherichia coli* has three SODs, namely, the Fe-SOD and Mn-SOD in the cytoplasm, and the Cu/Zn-SOD in the periplasm [13]. On the other hand, *H. pylori* produces only a single iron cofactored SOD (*Hp*SOD) encoded by the *sodB* gene (fig. 1) [14, 15]. Recently, the structure of *Hp*SOD was clarified, and it is now known to exist as a dimer composed of two identical subunits with one iron-ion monomer [16]. The protein shares 53% sequence identity with the corresponding enzyme from *E. coli*. *Hp*SOD shows significant differences from other SODs, the most characteristic being an extended C-terminal tail [16], although the role of this structural characteristic is not yet clarified. Seyler et al. [17] reported that *Hp*SOD deletion mutants showed increased O_2 sensitivity, in terms of both growth and viability. Furthermore, it was reported that the *Hp*SOD deletion mutants showed a reduced ability for colonization of mice [17]. From these reports, it is conceivable that *Hp*SOD is a virulence factor which affects the stomach-colonization ability of the bacteria. The *Hp*SOD expression was reported to be mediated by the ferric uptake regulator (Fur) protein [18]. Namely, the operator sequence of Fur, called the Fur-Box, was located directly upstream of the *sodB* gene at positions -5 to -47 from the transcriptional start site [18]. It is known that the Fur protein acts as an iron-dependent transcriptional repressor of *H. pylori* [19]. The iron-free form of Fur (apo-Fur) binds to the Fur-Box in the *sodB* promoter, which results in the suppression of *sodB* expression. In contrast, iron-binding Fur dissociates from the Fur-Box, derepressing the expression of *sodB* [18]. Recently, we reported that amino-acid mutations of Fur (C78Y, P114S and N118H; mutant-type Fur) induce the expression of *sodB* mRNA and *Hp*SOD activity, since the affinity of mutant-type Fur to the Fur-Box is significantly decreased [20]. Furthermore, this is a novel mechanism of development of resistance to metronidazole, which exerts its antibacterial activity via generation of superoxide radicals [20].

Catalase of *H. pylori*

KatA is a ubiquitous enzyme that catalyzes the decomposition of H_2O_2 into water and oxygen to protect cells from the damaging effects of H_2O_2. H_2O_2 has the potential to cause widespread cellular damage. Especially the rapid formation of other ROS from H_2O_2, including hydroxyl radicals, is of great significance. H_2O_2 reacts with intracellular free iron, in the well-characterized Fenton reaction, to produce hydroxyl radicals. H_2O_2 is the most toxic of all ROS; it can produce severe damage of many biomolecules, including DNA. While no enzymatic protection exists against the hydroxyl radicals, KatA is responsible for the removal of both exogenously produced and endogenously produced H_2O_2. KatA is expressed at high levels in *H. pylori*, accounting for approximately 1% of the total cell protein (fig. 1) [21]. KatA deletion mutant strains are viable in vitro under standard growth conditions, indicating that endogenously generated H_2O_2 is not significant in *H. pylori* [22, 23]. However, KatA is important for the survival of the bacterium in the presence of extracellular ROS produced by professional phagocytes and macrophage phagosomes [7, 24]. In addition, Harris et al. [25, 26] reported that while wild-type *H. pylori* can withstand exposure to ~100 mM of H_2O_2, the *katA*-deleted strains die within a few minutes under these conditions, suggesting that KatA-associated protein (KapA) which is encoded by a gene downstream of *katA* is involved in H_2O_2 resistance (fig. 1) [25]. Although deletion of *kapA* does not affect KatA activity, it increases the sensitivity to H_2O_2 [25]. In addition, *KapA* is considered to be a gene specific to *H. pylori*, as no homologue has been found in any other species.

A putative Fur-Box upstream of the coding region was identified in addition to that for *sodB*, suggesting a role of Fur in the regulation of *katA* expression [22, 23]. In contrast, it has been reported that Fur exhibits a weak affinity to the Fur-Box of *katA*, and furthermore, that little transcriptional difference of *katA* expression exists between the wild-type *H. pylori* and the Fur-deletion mutant strains [19]. Therefore, the potential for Fur to influence KatA expression in *H. pylori* warrants further investigation.

Other Antioxidant Ability-Related Proteins of *H. pylori*

The *H. pylori* genome contains a gene encoding a 26-kDa alkyl hydroperoxide reductase (AhpC) protein which catalyzes the reduction of hydrogen peroxide, peroxynitrite and organic hydroperoxides to their corresponding alcohols [27]. *H. pylori* AhpC exhibits broad specificity for the hydroperoxide substrate. It has been reported that the *ahpC*-deletion mutants are highly sensitive to oxidative stress and mutate more frequently than the wild-type *H. pylori* [28]. Furthermore, Olczak et al. [29] reported that the *ahpC*-deletion mutants were unable to colonize the stomachs of mice.

NapA of *H. pylori* is encoded by the gene HP0243 in the *H. pylori* genome. NapA was first identified as a factor that mediates the activity of the neutrophil activation protein [30]. By contrast, it has been reported that NapA is a member of the Dps family of proteins which function in iron binding and DNA protection [31]. Recently, Cooksley et al. [32] reported that NapA contributes to the ability of *H. pylori* to survive under oxidative conditions and binds to the DNA to protect the bacterium from oxidative damage (fig. 1). In addition, it is known that *napA*-deleted strains have significantly decreased ability to survive in mice as compared to the wild-type *H. pylori* [33].

Regulation System of Antioxidant Proteins

For chronic infection of the stomach, it is thought that the effective expression ability of these antioxidants is necessary for resistance against ROS. However, *H. pylori* possesses a surprisingly low number of regulators potentially involved in regulating the expression of antioxidant proteins; no homologues of the oxidative stress regulatory proteins present in other bacteria, such as OxyR, SoxR, SoxS, RpoS, LexA and PerR, have been found in the *H. pylori* genome. Since gene expression profiling using a microarray for *H. pylori* in response to oxidative stress has not yet been conducted, the regulation system of antioxidant proteins of *H. pylori* is not yet fully understood and warrants further investigation. Recently, it has been reported that the Fur proteins participating in iron homeostasis in the bacterial cell participates in the expression of *sodB*, *katA* and *napA* [18, 22, 23, 32]. Furthermore, it is well known that *fur* deletion mutant strains show altered expression of proteins with diverse functions, including energy metabolism, transcription and translation, biosynthesis of amino acids and nucleotides and production of the cell envelope [34, 35]. From these reports, it is conceivable that Fur is a master transcriptional regulator for the biological activity of *H. pylori,* and that therefore, the regulation system of antioxidant proteins by Fur warrants further investigation.

Host Side Oxidative Stress

Among the potential toxic factors involved in *H. pylori*-induced gastric injury are oxygen radicals which are released from activated neutrophils. Neutrophil infiltration of the gastric mucosa is reported as the initial pathological abnormality in *H. pylori*-associated gastritis and remains a hallmark of active infection. In response to the activation of neutrophils, NADPH oxidase in the cell membranes becomes activated, and an electron transfer takes place from NADPH in the cells to oxygen inside and outside the cells, and the oxygen molecules that receive an electron become superoxide radicals (O_2^-), which is rapidly converted to H_2O_2 by spontaneous dismutation or enzymatic SOD, and to hydroxyl radicals, which are formed nonenzymatically in the presence of Fe^{2+}. In neutrophils, myeloperoxidase also results in the formation of the potent oxidant, the hypochlorite anion (OCN^-), from H_2O_2 in the presence of Cl^-. This hypochlorite anion reacts with ammonia derived from urea produced by the action of *H. pylori*-associated urease, to yield monochloramine, which enhances the mucosal cytotoxicity based on its lipophilic property, and freely penetrates biological membranes to oxidize intracellular components [36]. On the other hand, ammonia itself also induces vacuolation of the gastric epithelial cells, necrosis of parietal cells and apoptosis of chief cells [37]. It is suggested that *H. pylori* may produce substances that degrade mucus and injure epithelial cells, thereby reducing the resistance of the mucosa to acid-related injury.

ROS production, as determined by chemiluminescence assay, has been shown to be enhanced in the gastric mucosa of *H. pylori*-infected patients with gastric ulcers. We previously showed that ROS production in the gastric mucosa is enhanced in infection caused by the *cagA*-positive *H. pylori* species, with extensive accumulation of neutrophils, in patients with gastric ulcer [38]. These findings were reconfirmed by a study in patients with chronic gastritis in the absence of peptic ulcer [39].

The excessive ROS production induces oxidative stress in the gastric mucosa, and may damage cellular components. Some of the gastroprotective agents that could mitigate the oxidative stress, such as rebamipide and polaprezinc, have been demonstrated to have properties, both in vitro and in vivo [40–42]. Sasazuki et al. [43] evaluated the effect of vitamin C supplementation on the serum levels of ROS among subjects with chronic gastritis, and concluded that vitamin C reduced the mucosal oxidative stress among subjects with atrophic gastritis.

Oxidative Stress and Apoptosis

H. pylori exerts much of its pathogenicity by inducing apoptosis and DNA damage of the host gastric epithelial cells. We previously demonstrated that monochloramine induced an increase in the chromatin condensation as well as in the cytoplasmic mono- and oligonucleosomes, one of the markers of apoptosis, in the gastric epithelial cell lines [44, 45]. On the other hand, polyamines are present in abundance in epithelial cells, and produce H_2O_2 when oxidized by the inducible spermine oxidase SMO (PAOhl). Xu et al. [46] identified a pathway for oxidative stress-induced epithelial cell apoptosis and DNA damage occurring as a result of SMO (PAOhl) activation by *H. pylori* that may contribute to the pathogenesis of the infection and the development of gastric cancer.

Calvino-Fernandez et al. [47] demonstrated that superoxide acts on mitochondria to initiate apoptotic pathways, with these changes occurring in the presence of mitochondrial depolarization and other morphological and functional changes. Moreover, treatment of infected cells with vitamin E prevented increase in the intracellular ROS and mitochondrial damage, consistent with *H. pylori* triggering a mitochondrial ROS-mediated programmed cell death pathway.

Nitric Oxide-Mediated Injury of the Gastric Mucosa

Nitric oxide (NO) is a regulator of the gastric mucosal microcirculation under resting and stimulated conditions, and interacts with prostanoids and sensory neuropeptides to maintain the gastric mucosal integrity. For the mucosal protective action of NO, regulation of gastric mucosal blood flow is considered to be of major importance. Other defense mechanisms induced by NO are stimulation of mucus secretion in the epithelial cells, inhibition of the aggregation and adhesion of platelets and inhibition of the adherence of neutrophils to the endothelium and their emigration to blood vessels and scavenging of O_2^-, which cause mucosal damage. However, bicarbonate secretion, one of the defensive factors in the gastroduodenal mucosa, has been shown to be increased by NO synthase inhibitors, which is contradictory to the mucosal protective role of NO. On the other hand, O_2^- and NO rapidly react to form peroxynitrite, which decomposes and generates strong oxidant molecules. Large amounts of NO produced by inducible NO synthase (iNOS) activated by *H. pylori* in the gastric epithelial AGS cells contribute to apoptotic cell death. Studies clearly demonstrate the involvement of oxidative stress in *H. pylori*-stimulated expression of iNOS and apoptosis in gastric AGS cells [48]. On the other hand, Miyazawa et al. [49] reported suppressed apoptosis in the inflamed gastric mucosa of *H. pylori*-colonized iNOS knockout mice, and suggested that iNOS may play an important role in promoting apoptosis in *H. pylori*-associated gastritis, and that persistent inflammation without apoptosis, such as in iNOS knockout mice with *H. pylori* infection, may be linked to preneoplastic transformation. These contentions may reflect the biphasic aspects of the roles of iNOS-derived NO in the regulation of cell turnover.

Conclusion

The antioxidant ability of *H. pylori* is unique as compared with that of other bacteria: (a) *H. pylori* produces only a single iron-cofactored SOD encoded by the *sodB* gene; (b) *KapA* is considered to be a gene specific to *H. pylori* that is involved in H_2O_2 resistance; (c) no homologues of the oxidative stress regulator proteins in *H. pylori* have been found in other bacterial strains. The ability of these unique antioxidant systems of *H. pylori* may be important in the ability of *H. pylori* to produce chronic infection of the stomach.

On the other hand, reduced host defenses, including of the antioxidant capacity of the gastric mucosa, are major pathogenetic mechanisms underlying the development of gastric disorders associated with *H. pylori* infection. Inflammatory genes and related mediators stimulating cell proliferation and defensive molecular chaperones are induced by *H. pylori* in the gastric epithelial cells, which are believed to be mediated by oxygen-derived free radicals. Thus, gastric mucosal oxidative stress occurs as a result of factors operating from both the bacterial and host sides [50]. A broad knowledge of the source of oxygen radicals, mechanisms of injury initiation and biochemical regulation of antioxidant defense systems would be expected to provide further insight into the development of novel therapeutic strategies for the regulation of *H. pylori*-associated gastroduodenal diseases.

References

1 Suzuki H, Marshall BJ, Hibi T: Overview: *Helicobacter pylori* and Extragastric Disease. Int J Hematol 2006;84:291–300.

2 Suzuki H, Hibi T, Marshall BJ: *Helicobacter pylori*: present status and future prospects in Japan. J Gastroenterol 2007;42:1–15.

3 Backert S, Schwarz T, Miehlke S, Kirsch C, Sommer C, Kwok T, Gerhard M, Goebel UB, Lehn N, Koenig W, Meyer TF: Functional analysis of the cag pathogenicity island in *Helicobacter pylori* isolates from patients with gastritis, peptic ulcer, and gastric cancer. Infect Immun 2004;72:1043–1056.

4 Ding SZ, Minohara Y, Fan XJ, Wang J, Reyes VE, Patel J, Dirden-Kramer B, Boldogh I, Ernst PB, Crowe SE: *Helicobacter pylori* infection induces oxidative stress and programmed cell death in human gastric epithelial cells. Infect Immun 2007;75:4030–4039.

5 Barnard FM, Loughlin MF, Fainberg HP, Messenger MP, Ussery DW, Williams P, Jenks PJ: Global regulation of virulence and the stress response by CsrA in the highly adapted human gastric pathogen *Helicobacter pylori*. Mol Microbiol 2004;51:15–32.

6 Bagchi D, Bhattacharya G, Stohs SJ: Production of reactive oxygen species by gastric cells in association with *Helicobacter pylori*. Free Radic Res 1996 24:439–450.

7 Ramarao N, Gray-Owen SD, Meyer TF: *Helicobacter pylori* induces but survives the extracellular release of oxygen radicals from professional phagocytes using its catalase activity. Mol Microbiol 2000;38:103–113.

8 Davies GR, Simmonds NJ, Stevens TR, Sheaff MT, Banatvala N, Laurenson IF, Blake DR, Rampton DS: *Helicobacter pylori* stimulates antral mucosal reactive oxygen metabolite production in vivo. Gut 1994;35:179–185.

9 Baik SC, Youn HS,Chung MH, Lee WK, Cho MJ, Ko GH, Park CK, Kasai H, Rhee KH: Increased oxidative DNA damage in *Helicobacter pylori*-infected human gastric mucosa. Cancer Res 1996;56:1279–1282.
10 Storz G, Imlay JA: Oxidative stress. Curr Opin Microbiol 1999;2:188–194.
11 Wuerges J, Lee JW, Yim YI, Yim HS, Kang SO, Djinovic Carugo K: Crystal structure of nickel-containing superoxide dismutase reveals another type of active site. Proc Natl Acad Sci U S A 2004;101:8569–8574.
12 Barondeau DP, Kassmann CJ,Bruns CK, Tainer JA, Getzoff ED: Nickel superoxide dismutase structure and mechanism. Biochemistry 2004;43:8038–8047.
13 Benov LT, Fridovich I: Escherichia coli expresses a copper- and zinc-containing superoxide dismutase. J Biol Chem 1994;269:25310–25314.
14 Spiegelhalder C, Gerstenecker B, Kersten A, Schiltz E, Kist M: Purification of *Helicobacter pylori* superoxide dismutase and cloning and sequencing of the gene. Infect Immun 1993;61:5315–5325.
15 Pesci EC, Pickett CL: Genetic organization and enzymatic activity of a superoxide dismutase from the microaerophilic human pathogen, *Helicobacter pylori*. Gene 1994;143:111–116.
16 Esposito L, Seydel A, Aiello R, Sorrentino G, Cendron L, Zanotti G, Zagari A: The crystal structure of the superoxide dismutase from *Helicobacter pylori* reveals a structured C-terminal extension. Biochim Biophys Acta 2008;1784:1601–1606.
17 Seyler RW Jr, Olson JW, Maier RJ: Superoxide dismutase-deficient mutants of *Helicobacter pylori* are hypersensitive to oxidative stress and defective in host colonization. Infect Immun 2001;69:4034–4040.
18 Ernst FD, Homuth G, Stoof J, Mader U, Waidner B, Kuipers EJ, Kist M, Kusters JG, Bereswill S, van Vliet AH: Iron-responsive regulation of the *Helicobacter pylori* iron-cofactored superoxide dismutase SodB is mediated by Fur. J Bacteriol 2005;187:3687–3692.
19 Delany I, Spohn G, Rappuoli R, Scarlato V: The Fur repressor controls transcription of iron-activated and -repressed genes in *Helicobacter pylori*. Mol Microbiol 2001;42:1297–1309.
20 Tsugawa H, Suzuki H, Satoh K, Hirata K, Matsuzaki J, Saito Y, Suematsu M, Hibi T: Two amino acids mutation of Ferric uptake regulator (Fur) determines *Helicobacter pylori* resistance to metronidazole. Antioxid. Redox Signal. 2010 Jun 2, Epub, ahead of print.
21 Hazell SL, Evans DJ Jr, Graham DY: *Helicobacter pylori* catalase. J Gen Microbiol 1991;137:57–61.
22 Odenbreit S, Wieland B, Haas R: Cloning and genetic characterization of *Helicobacter pylori* catalase and construction of a catalase-deficient mutant strain. J Bacteriol 1996;178:6960–6967.
23 Manos J, Kolesnikow T, Hazell SL: An investigation of the molecular basis of the spontaneous occurrence of a catalase-negative phenotype in *Helicobacter pylori*. Helicobacter 1998;3:28–38.
24 Basu M, Czinn SJ, Blanchard TG: Absence of catalase reduces long-term survival of *Helicobacter pylori* in macrophage phagosomes. Helicobacter 2004;9:211–216.
25 Harris AG, Hinds FE, Beckhouse AG, Kolesnikow T, Hazell SL: Resistance to hydrogen peroxide in *Helicobacter pylori*: role of catalase (KatA) and Fur, and functional analysis of a novel gene product designated 'KatA-associated protein', KapA (HP0874). Microbiology 2002;148:3813–3825.
26 Harris AG, Wilson JE, Danon SJ, Dixon MF, Donegan K, Hazell SL: Catalase (KatA) and KatA-associated protein (KapA) are essential to persistent colonization in the *Helicobacter pylori* SS1 mouse model. Microbiology 2003;149:665–672.
27 Wood ZA, Schroder E, Robin Harris J, Poole LB: Structure, mechanism and regulation of peroxiredoxins. Trends Biochem Sci 2003;28:32–40.
28 Olczak AA, Olson JW, Maier RJ: Oxidative-stress resistance mutants of *Helicobacter pylori*. J Bacteriol 2002;184:3186–3193.
29 Olczak AA, Seyler RW Jr, Olson JW, Maier RJ: Association of *Helicobacter pylori* antioxidant activities with host colonization proficiency. Infect Immun 2003;71:580–583.
30 Evans DJ Jr, Evans DG, Takemura T, Nakano H, Lampert HC, Graham DY, Granger DN, Kvietys PR: Characterization of a *Helicobacter pylori* neutrophil-activating protein. Infect Immun 1995;63:2213–2220.
31 Grant RA, Filman DJ, Finkel SE, Kolter R, Hogle JM: The crystal structure of Dps, a ferritin homolog that binds and protects DNA. Nat Struct Biol 1998;5:294–303.
32 Cooksley C, Jenks PJ, Green A, Cockayne A, Logan RP, Hardie KR: NapA protects *Helicobacter pylori* from oxidative stress damage, and its production is influenced by the ferric uptake regulator. J Med Microbiol 2003;52:461–469.
33 Wang G, Hong Y, Olczak A, Maier SE, Maier RJ: Dual Roles of *Helicobacter pylori* NapA in inducing and combating oxidative stress. Infect Immun 2006;74:6839–6846.
34 Lee HW, Choe YH, Kim DK, Jung SY, Lee NG: Proteomic analysis of a ferric uptake regulator mutant of *Helicobacter pylori*: regulation of *Helicobacter pylori* gene expression by ferric uptake regulator and iron. Proteomics 2004;4:2014–2027.

35 Ernst FD, Bereswill S, Waidner B, Stoof J, Mader U, Kusters JG, Kuipers EJ, Kist M, van Vliet AH, Homuth G: Transcriptional profiling of *Helicobacter pylori* Fur- and iron-regulated gene expression. Microbiology 2005;151:533–546.

36 Suzuki M, Miura S, Suematsu M, Fukumura D, Kurose I, Suzuki H, Kai A, Kudoh Y, Ohashi M, Tsuchiya M: *Helicobacter pylori*-associated ammonia production enhances neutrophil-dependent gastric mucosal cell injury. Am J Physiol 1992;263: G719—G725.

37 Hagen SJ, Takahashi S, Jansons R: Role of vacuolation in the death of gastric epithelial cells. Am J Physiol 1997;272:C48–C58.

38 Suzuki H, Suzuki M, Mori M, Kitahora T, Yokoyama H, Miura S, Hibi T, Ishii H: Augmented levels of gastric mucosal leukocyte activation by infection with cagA gene positive *Helicobacter pylori*. J Gastroenterol Hepatol 1998;13:294–300.

39 Danese S, Cremonini F, Armuzzi A, Candelli M, Papa A, Ojetti V, Pastorelli A, Di Caro S, Zannoni G, De Sole P, Gasbarrini G, Gasbarrini A: *Helicobacter pylori* CagA-positive strains affect oxygen free radicals generation by gastric mucosa. Scand J Gastroenterol 2001;36:247–250.

40 Suzuki M, Miura S, Mori M, Kai A, Suzuki H, Fukumura D, Suematsu M, Tsuchiya M: Rebamipide, a novel antiulcer agent, attenuates *Helicobacter pylori* induced gastric mucosal cell injury associated with neutrophil derived oxidants. Gut 1994;35:1375–1378.

41 Suzuki H, Mori M, Seto K, Miyazawa M, Kai A, Suematsu M, Yoneta T, Miura S, Ishii H: Polaprezinc attenuates the *Helicobacter pylori*-induced gastric mucosal leucocyte activation in Mongolian gerbils – a study using intravital videomicroscopy. Aliment Pharmacol Ther 2001;15:715–725.

42 Nishizawa T, Suzuki H, Nakagawa I, Minegishi Y, Masaoka T, Iwasaki E, Hibi T: Rebamipide-promoted restoration of gastric mucosal sonic hedgehog expression after early *Helicobacter pylori* eradication. Digestion 2009;79:259–262.

43 Sasazuki S, Hayashi T, Nakachi K, Sasaki S, Tsubono Y, Okubo S, Hayashi M, Tsugane S: Protective effect of vitamin C on oxidative stress: a randomized controlled trial. Int J Vitam Nutr Res 2008;78:121–128.

44 Suzuki H, Seto K, Mori M, Suzuki M, Miura S, Ishii H: Monochloramine induced DNA fragmentation in gastric cell line MKN45. Am J Physiol 1998; 275:G712–G716.

45 Suzuki H, Mori M, Suzuki M, Sakurai K, Miura S, Ishii H: Extensive DNA damage induced by monochloramine in gastric cells. Cancer Lett 1997;115: 243–248.

46 Xu H, Chaturvedi R, Cheng Y, Bussiere FI, Asim M, Yao MD, Potosky D, Meltzer SJ, Rhee JG, Kim SS,Moss SF, Hacker A, Wang Y, Casero RA Jr, Wilson KT: Spermine oxidation induced by *Helicobacter pylori* results in apoptosis and DNA damage: implications for gastric carcinogenesis. Cancer Res 2004; 64:8521–8525.

47 Calvino-Fernandez M, Benito-Martinez S, Parra-Cid T: Oxidative stress by *Helicobacter pylori* causes apoptosis through mitochondrial pathway in gastric epithelial cells. Apoptosis 2008;13:1267–1280.

48 Lim JW, Kim H, Kim KH: NF-kappaB, inducible nitric oxide synthase and apoptosis by *Helicobacter pylori* infection. Free Radic Biol Med 2001;31:355–366.

49 Miyazawa M, Suzuki H, Masaoka T, Kai A, Suematsu M, Nagata H, Miura S, Ishii H: Suppressed apoptosis in the inflamed gastric mucosa of *Helicobacter pylori*-colonized iNOS-knockout mice. Free Radic Biol Med 2003;34:1621–1630.

50 Suzuki H, Suzuki M, Imaeda H, Hibi T: *Helicobacter pylori* and Microcirculation. Microcirculation 2009; 16:547–558.

Hidekazu Suzuki, MD, PhD
Division of Gastroenterology and Hepatology
Department of Internal Medicine, Keio University School of Medicine
35 Shinanomachi, Shinjuku-ku
Tokyo 160-8582, Japan
Tel. +81 3 5363 3914, Fax +81 3 5363 3967, E-Mail hsuzuki@sc.itc.keio.ac.jp

Naito Y, Suematsu M, Yoshikawa T (eds): Free Radical Biology in Digestive Diseases.
Front Gastrointest Res. Basel, Karger, 2011, vol 29, pp 121–127

Free Radical and Nonsteroidal Anti-Inflammatory Drug-Induced Small Intestinal Injury

Osamu Handa · Yuji Naito

Department of Molecular Gastroenterology and Hepatology, Kyoto Prefectural University of Medicine, Kyoto, Japan

Abstract

The small intestine is a difficult organ for gastroenterologists to access and study since it is located far from both the mouth and the anus. This, in turn, makes it very difficult to diagnose or treat diseases of the small intestine. However, recent technological developments, including video capsule endoscopy and balloon endoscopy, have enabled us to look inside the dark continent of the digestive tract. As a result, various diseases of the small intestine have been revealed. Of them, nonsteroidal anti-inflammatory drug (NSAID)-induced mucosal damage of the small intestine is of great interest to gastroenterologists since the incidence of this disease is considerably higher than previously believed. At its worst, NSAID-induced mucosal damage results in life-threatening bleeding from the small intestine, even after administration of gastroprotective or antigastric acid secretion drugs. Therefore, identification of a drug that effectively protects the small intestine from mucosal damage is urgently required. Moreover, there are currently no effective drugs for the clinical treatment of NSAID-induced mucosal damage of the small intestine. In this paper, we summarize the molecular mechanism by which NSAIDs damage the small intestine and also describe the involvement of reactive oxygen species.

The adverse effects of nonsteroidal anti-inflammatory drugs (NSAIDs) on the stomach and duodenum have been intensively investigated, but their adverse effects on the small intestine have not been sufficiently studied.

Recent advances in endoscopic technology, including balloon endoscopy, such as single or double balloon endoscopy (DBE), and video capsule endoscopy, have allowed observation of the small intestinal mucosa in more detail. Consequently, it is now clear that NSAIDs cause damage not only to gastric and duodenal mucosa, but also to the small intestinal mucosa with relatively high incidence.

The adverse effects of NSAIDs have been investigated using video capsule endoscopy. Graham et al. [1] found that 71% of arthritis patients who take various NSAIDs for more than 3 months exhibit intestinal mucosal damage, compared to 10% of the individuals who took either acetaminophen alone or nothing as a control group. Maiden et al. [2] reported that 68% of 40 healthy volunteers exhibited intestinal mucosal damage after 14 days on diclofenac (a traditional NSAID) and omeprazole (a proton pump inhibitor). Using DBE, Matsumoto et al. [3] found that 51% of 61 patients taking NSAIDs for a month prior to DBE exhibited nonspecific mucosal breaks, whereas 5% of 600 patients in the control group who did not use NSAIDs exhibited mucosal breaks. As our society ages, the number of elderly individuals and patients who frequently take NSAIDs are increasing. Therefore, establishing a method for preventing or treating NSAID-induced small intestinal injury is urgently required.

In this paper, we describe the molecular mechanism by which NSAIDs cause intestinal mucosal damage and also describe the involvement of reactive oxygen species (ROS). By clarifying the pathological mechanism, new strategies for preventing or treating NSAID-induced mucosal damage to the small intestine can be developed.

The Molecular Mechanism of NSAID-Induced Intestinal Damage

Multiple factors may be involved in the molecular mechanism of NSAID-induced intestinal damage such as the host macrobiotic condition, food intake, medication, pre-existing diseases, and age. However, cyclooxygenase (COX) appears to be a primary culprit in the pathogenesis of NSAID-induced intestinal damage. Therefore, in this section we classify the molecular mechanism into two categories: a COX-dependent mechanism and a COX-independent mechanism.

COX-Dependent Mechanism

Prostaglandins (PGs) regulate gastrointestinal blood flow, mucus secretion and intestinal motility, thereby providing mucosal protective functions. COX is a well-known enzyme that mediates the synthesis of PGs.

The effects of NSAIDs depend on their inhibition of COX activity at inflammation sites where NSAIDs exhibit analgesic, antipyretic and anti-inflammatory effects. Since the inhibition of PG production is thought to be the major cause of NSAID-induced small intestinal injury, the effects of COX inhibition have been extensively investigated.

COX has two subtypes, COX-1 and COX-2. COX-1 mediated-PG synthesis is especially important for maintaining mucosal homeostasis. The selective inhibition of COX-1 has been shown to incrementally increase mucosal permeability, aid the invasion of intestinal bacteria into the mucosa, reduce intestinal mucosal microcirculation

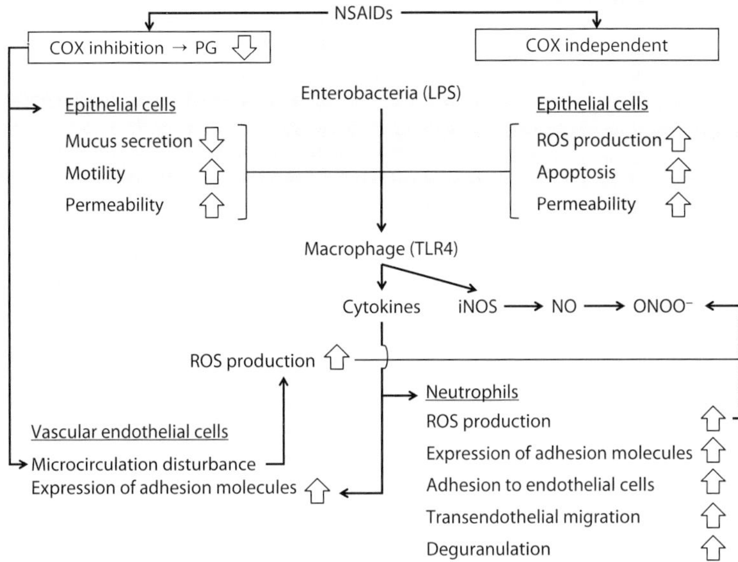

Fig. 1. Working hypothesis of NSAID-induced small intestinal injury.

[4], and induce the synthesis of nitric oxide synthase (iNOS). In contrast, COX-2 plays a role in the production of PGE2 to protect the intestinal mucosa [5]. Consequently, COX-2 selective NSAIDs have been developed, and many researchers have reported that COX-2 selective NSAIDs cause less damage to the small intestine than non-COX-selective NSAIDs [6]. However, two recent reports have indicated that there are no differences in the incidence of small intestinal mucosal damage between COX-2 selective NSAIDs and traditional non-COX-selective NSAIDs [7, 8]. This issue, therefore, requires further investigation.

COX-Independent Mechanism

The 'three hit theory' is currently the most widely-accepted COX-independent pathogenic mechanism by which NSAIDs induce intestinal mucosal damage [9].

First, NSAIDs lyse the phospholipid membrane on the mucosal surface and directly injure the mitochondria in the intestinal epithelial cells. Subsequently, damaged mitochondrial function results in calcium efflux, excessive free radical formation, reduced intercellular connections, and increased permeability of the intestinal mucosa. Finally, bile acids, proteases, intestinal bacteria and toxins gain entry through increased intercellular junctions that ultimately results in mucosal injury. As Bjarnason et al. [10] concluded, the primary adverse effect of NSAIDs is the incremental increase in the mucosal permeability of the small intestine, and the secondary adverse effect is inflammation or ulcer formation caused by neutrophils activated by bile acids or intestinal bacteria.

Enterohepatic circulation of NSAIDs has been reported to be involved in their toxicity to the small intestine. For example, enterohepatic circulation-independent drugs such as sulindac have lower toxicity to small intestinal mucosa [11], although this should be confirmed by further investigation.

Another important factor is intestinal bacteria. Several lines of evidence indicate the importance of intestinal bacteria in NSAID-induced small intestinal damage. For example, indomethacin (a traditional NSAID) does not cause mucosal damage in germ-free rats [12, 13] and some antibiotics such as metronidazole [10] and ampicillin [14] or probiotics (e.g. *Lactobacillus* casi [15]) suppress indomethacin-induced intestinal mucosal damage. These results suggest a pivotal involvement of intestinal bacteria in NSAID-induced mucosal damage.

The involvement of Gram-negative bacteria has also been reported. Toll-like receptor 4 (TLR4) recognizes the lipopolysaccharides of Gram-negative bacteria and mediates inflammatory activation, and it has been shown to be actively involved in mucosal damage. Watanabe et al. [16] showed that indomethacin- and diclofenac-induced intestinal mucosal damage was significantly suppressed in TLR4 mutant mice. Since the expression of TLR4 on small intestinal epithelial cells is less than that on intestinal mucosal macrophages, their results might indicate that the role of Gram-negative bacteria in mucosal damage is primarily due to their effect on macrophages rather than on epithelial cells [15].

An interesting report has recently been published by Pilotto et al. [17]. They suggested that genetic differences might be involved in increased susceptibility of the small intestine to injury: polymorphism of CYP2C, an enzyme that metabolizes NSAIDs, modified the risk of NSAID-related gastroduodenal bleeding. This finding may explain why only some patients taking NSAIDs exhibit damage to the small intestine; further studies are clearly required.

Reactive Oxygen Species and NSAID-Induced Small Intestinal Damage

ROS have been reported to play an important role in the pathogenesis of NSAID-induced small intestinal damage [18]. There are three possible sources of ROS at the inflamed sites of NSAID-injured small intestine: epithelial cells, vascular endothelial cells, and inflammatory cells.

Small intestinal epithelial cells are the first line of defense against NSAIDs; however, the role of ROS in small intestinal epithelial cells has not been explored. Recently, we reported that indomethacin induces ROS production in the mitochondria of small intestinal epithelial cells and the apoptosis of these cells. This cell injury might lead to intestinal mucosal damage that is independent of PG suppression [19, 20]. The produced ROS might modify intracellular proteins and lipids, or interact with nitric oxide (NO) to cause further injury, as described below.

The importance of ROS production in vascular endothelial cells in NSAID enteropathy is presently not clear. Anthony et al. [21] reported that in the rat jejunum, poorly vascularized sites along the mesenteric margin were highly susceptible to indomethacin-induced ischemia-reperfusion injury. This might be a clue to clarifying the importance of ROS production in vascular endothelial cells in NSAID enteropathy since ROS production in vascular endothelial cells has been shown to be important for the early phase of anoxia-reoxygenation injury [22].

Marked accumulation of neutrophils and oxygen-derived free radical production has been reported in the small intestinal mucosa of indomethacin-administered rat [23]. The importance of neutrophil activation was shown in a study using gp91(phox$^{-/-}$) mice, which are less susceptible to NSAID injury [24]. Moreover, NO production by other inflammatory cells might be important for NSAID enteropathy. Constitutive NO synthase-derived NO has been shown to maintain mucosal homeostasis by adjusting mucosal blood flow and mucus secretion. On the other hand, iNOS-derived NO has been shown to cause mucosal damage in the small intestine of indomethacin-administered rats [25], and the produced NO interacts with ROS to form peroxynitrite, which may be involved in ulcer formation in the small intestine [26]. Takeuchi et al. [25] found that administration of ampicillin reduced the expression of iNOS in intestinal mucosa, suggesting the pivotal role of intestinal bacteria in NSAID enteropathy.

Conclusions

Many NSAID users need a safe mucoprotective drug, making it imperative that the precise mechanism by which NSAIDs induce small intestinal damage be clarified as soon as possible. Understanding the mechanism will provide useful insights for developing novel strategies for preventing or treating NSAID-induced small intestinal damage.

References

1 Graham DY, Opekun AR, Willingham FF, Qureshi WA: Visible small-intestinal mucosal injury in chronic NSAID users. Clin Gastroenterol Hepatol 2005;3:55–59.

2 Maiden L, Thjodleifsson B, Theodors A, Gonzalez J, Bjarnason I: A quantitative analysis of NSAID-induced small bowel pathology by capsule enteroscopy. Gastroenterology 2005;128:1172–1178.

3 Matsumoto T, Kudo T, Esaki M, Yano T, Yamamoto H, Sakamoto C, Goto H, Nakase H, Tanaka S, Matsui T, Sugano K, Iida M: Prevalence of non-steroidal anti-inflammatory drug-induced enteropathy determined by double-balloon endoscopy: a Japanese multicenter study. Scand J Gastroenterol 2008; 43:490–496.

4 Kelly DA, Piasecki D, Anthony A, Dhillon AP, Pounder RE, Wakefield AJ: Focal reduction of villous blood flow in early indomethacin enteropathy: a dynamic vascular study in the rat. Gut 1998;42:366–373.

5 Tanaka A, Hase S, Miyazawa T, Ohno R, Takeuchi K: Role of cyclooxygenase (COX)-1 and COX-2 inhibition in nonsteroidal anti-inflammatory drug-induced intestinal damage in rats: relation to various pathogenic events. J Pharmacol Exp Ther 2002;303:1248–1254.

6 Goldstein JL, Eisen GM, Lewis B, Gralnek IM, Zlotnick S, Fort JG: Video capsule endoscopy to prospectively assess small bowel injury with celecoxib, naproxen plus omeprazole, and placebo. Clin Gastroenterol Hepatol 2005;3:133–141.

7 Sugimori S, Watanabe T, Tabuchi M, Kameda N, Machida H, Okazaki H, Tanigawa T, Yamagami H, Shiba M, Watanabe K, Tominaga K, Fujiwara Y, Oshitani N, Koike T, Higuchi K, Arakawa T: Evaluation of small bowel injury in patients with rheumatoid arthritis by capsule endoscopy: effects of anti-rheumatoid arthritis drugs. Digestion 2008;78:208–213.

8 Maiden L, Thjodleifsson B, Seigal A, Bjarnason II, Scott D, Birgisson S, Bjarnason I: Long-term effects of nonsteroidal anti-inflammatory drugs and cyclooxygenase-2 selective agents on the small bowel: a cross-sectional capsule enteroscopy study. Clin Gastroenterol Hepatol 2007;5:1040–1045.

9 Bjarnason I, Hayllar J, MacPherson AJ, Russell AS: Side effects of nonsteroidal anti-inflammatory drugs on the small and large intestine in humans. Gastroenterology 1993;104:1832–1847.

10 Bjarnason I, Hayllar J, Smethurst P, Price A, Gumpel MJ: Metronidazole reduces intestinal inflammation and blood loss in non-steroidal anti-inflammatory drug induced enteropathy. Gut 1992;33:1204–1208.

11 Smale S, Tibble J, Sigthorsson G, Bjarnason I: Epidemiology and differential diagnosis of NSAID-induced injury to the mucosa of the small intestine. Best Pract Res Clin Gastroenterol 2001;15:723–738.

12 Robert A, Asano T: Resistance of germfree rats to indomethacin-induced intestinal lesions. Prostaglandins 1977;14:333–341.

13 Whittle BJ, Laszlo F, Evans SM, Moncada S: Induction of nitric oxide synthase and microvascular injury in the rat jejunum provoked by indomethacin. Br J Pharmacol 1995;116:2286–2290.

14 Konaka A, Kato S, Tanaka A, Kunikata T, Korolkiewicz R, Takeuchi K: Roles of enterobacteria, nitric oxide and neutrophil in pathogenesis of indomethacin-induced small intestinal lesions in rats. Pharmacol Res 1999;40:517–524.

15 Watanabe T, Nishio H, Tanigawa T, Yamagami H, Okazaki H, Watanabe K, Tominaga K, Fujiwara Y, Oshitani N, Asahara T, Nomoto K, Higuchi K, Takeuchi K, Arakawa T: Probiotic *Lactobacillus casei* strain Shirota prevents indomethacin-induced small intestinal injury: involvement of lactic acid. Am J Physiol Gastrointest Liver Physiol 2009;297:G506–G513.

16 Watanabe T, Higuchi K, Kobata A, Nishio H, Tanigawa T, Shiba M, Tominaga K, Fujiwara Y, Oshitani N, Asahara T, Nomoto K, Takeuchi K, Arakawa T: Non-steroidal anti-inflammatory drug-induced small intestinal damage is Toll-like receptor 4 dependent. Gut 2008;57:181–187.

17 Pilotto A, Seripa D, Franceschi M, Scarcelli D, Colaizzo D, Grandone E, Niro V, Andriulli A, Leandro G, Di Mario F, Dallapiccola B: Genetic susceptibility to nonsteroidal anti-inflammatory drug-related gastroduodenal bleeding: role of cytochrome P450 2C9 polymorphisms. Gastroenterology 2007;133:465–471.

18 Basivireddy J, Jacob M, Ramamoorthy P, Pulimood AB, Balasubramanian KA: Indomethacin-induced free radical-mediated changes in the intestinal brush border membranes. Biochem Pharmacol 2003;65:683–695.

19 Omatsu T, Naito Y, Handa O, Hayashi N, Mizushima K, Qin Y, Hirata I, Adachi S, Okayama T, Kishimoto E, Takagi T, Kokura S, Ichikawa H, Yoshikawa T: Involvement of reactive oxygen species in indomethacin-induced apoptosis of small intestinal epithelial cells. J Gastroenterol 2009;44(Suppl 19):30–34.

20 Omatsu T, Naito Y, Handa O, Mizushima K, Hayashi N, Qin Y, Harusato A, Hirata I, Kishimoto E, Okada H, Uchiyama K, Ishikawa T, Takagi T, Yagi N, Kokura S, Ichikawa H, Yoshikawa T: Reactive oxygen species-quenching and anti-apoptotic effect of polaprezinc on indomethacin-induced small intestinal epithelial cell injury. J Gastroenterol 2010;45:692–702.

21 Anthony A, Pounder RE, Dhillon AP, Wakefield AJ: Vascular anatomy defines sites of indomethacin induced jejunal ulceration along the mesenteric margin. Gut 1997;41:763–770.

22 Ichikawa H, Kokura S, Aw TY: Role of endothelial mitochondria in oxidant production and modulation of neutrophil adherence. J Vasc Res 2004;41:432–444.

23 Miura S, Suematsu M, Tanaka S, Nagata H, Houzawa S, Suzuki M, Kurose I, Serizawa H, Tsuchiya M: Microcirculatory disturbance in indomethacin-induced intestinal ulcer. Am J Physiol 1991;261:G213–G219.

24 Beck PL, Xavier R, Lu N, Nanda NN, Dinauer M, Podolsky DK, Seed B: Mechanisms of NSAID-induced gastrointestinal injury defined using mutant mice. Gastroenterology 2000;119:699–705.

25 Takeuchi K, Yokota A, Tanaka A, Takahira Y: Factors involved in upregulation of inducible nitric oxide synthase in rat small intestine following administration of nonsteroidal anti-inflammatory drugs. Dig Dis Sci 2006;51:1250–1259.

26 Bjarnason I, Takeuchi K, Simpson R:NSAIDs: the emperor's new dogma? Gut 2003;52:1376–1378.

Osamu Handa
Department of Molecular Gastroenterology and Hepatology, Kyoto Prefectural University of Medicine
465 Kajiicho, Kawaramachi-Hirokoji agaru, Kamigyo-ku
Kyoto 602-8566 (Japan)
Tel. +81 75 251 5519, Fax +81 75 251 0710, E-Mail handao@koto.kpu-m.ac.jp

Naito Y, Suematsu M, Yoshikawa T (eds): Free Radical Biology in Digestive Diseases.
Front Gastrointest Res. Basel, Karger, 2011, vol 29, pp 128–136

Free Radicals in Inflammatory Bowel Disease

Tomohisa Takagi · Yuji Naito · Toshikazu Yoshikawa

Molecular Gastroenterology and Hepatology, Graduate School of Medical Science, Kyoto Prefectural University of Medicine, Kyoto, Japan

Abstract

Inflammatory bowel disease (IBD), including ulcerative colitis and Crohn's disease, is a chronic and recurrent inflammatory disorder of the colon and rectum. While the precise pathogenesis of IBD remains unknown, oxidative stress caused by reactive oxygen species and reactive nitrogen species is believed to be an important factor involved in the onset and development of intestinal inflammation. Furthermore, a depressed antioxidant defense system is involved in the pathophysiology of IBD. In this review, we present what is currently known regarding the role of oxidative stress in IBD.

Inflammatory bowel disease (IBD), including ulcerative colitis (UC) and Crohn's disease (CD), is a chronic and recurrent intestinal inflammatory disorder. The incidence of IBD is rapidly increasing in Japan as well as in western countries [1]. While certain features of the disease suggest several possible causes, including familial, genetic, infectious, physiological, and immunological factors [2, 3], the precise pathogenesis of IBD remains unknown. Therefore, it is important to investigate the pathogenesis of IBD and to evaluate new anti-inflammatory strategies.

Recent evidence from clinical and basic science studies reveals that enhanced formation of reactive oxygen species (ROS) or reactive nitrogen species (RNS) contributes to the pathophysiology of IBD. In addition, antioxidant defense systems against excessive ROS and RNS are suppressed in the inflamed intestine of IBD patients. This article provides an overview of the recent advances concerning the role of ROS, RNS, and the antioxidant defense system in IBD.

Reactive Oxygen Species Production in Inflammatory Bowel Disease

Although IBD is a multifactorial disease, one of the most prominent histological features observed in IBD is neutrophil infiltration into the inflamed mucosa. This is especially true of UC, and it is well known that disease activity in UC is linked to an influx of neutrophils into the mucosa and subsequently into the intestinal lumen, resulting in the formation of so-called crypt abscesses. In addition, circulating activated neutrophils, a major source of inflammatory cytokines, are elevated in active UC. Several studies have shown that granulocyte/monocyte adsorptive apheresis therapy can induce remission, especially in patients with steroid-refractory or steroid-dependent moderate to severe UC [4, 5]. Activated neutrophils are known potential sources of free radical production. On the other hand, monocytes/macrophages play an important role in free radical production in CD.

The presence of ROS has been extensively studied in patients with IBD. Free radical production has been estimated by chemiluminescence and electron spin resonance spectroscopy. The chemiluminescence method was used to directly quantify the ROS level in colon biopsy specimens from the patients with UC or CD, and this showed that the ROS level was markedly increased in these mucosa samples as compared to normal mucosa. However, because ROS have short biological half-lives, their presence is generally measured indirectly by assessing the levels of oxidatively damaged molecules. This includes assessment of lipid peroxidation because unsaturated bonds in membrane phospholipids are major targets for free radical reactions and oxidative DNA damage. Although there are some conflicting reports, the production of excessively oxidized molecules appears to be upregulated in patients with IBD as compared to healthy control subjects in a variety of organic systems, including serum, urine, and biopsy specimens from the colonic mucosa [6, 7].

It was previously thought that the increase in NADPH oxidase due to neutrophil and macrophage invasion resulted in oxidative stress in the colon. Recent investigations have revealed the presence of six homologs of the cytochrome subunit of the phagocyte NADPH oxidase: Nox1, Nox3, Nox4, Nox5, Duox1, and Duox2. Together with the phagocyte NADPH oxidase (Nox2/gp91^{phox}), the homologs are now referred to as the Nox family of NADPH oxidases. These enzymes share the capacity to transport electrons across the plasma membrane and to generate superoxide (O_2^-) and other downstream ROS. After the discovery of Nox homologs, it became clear that Nox1 is highly expressed in the distal colon epithelium [8, 9]. Although the function of Nox1 in the distal gastrointestinal tract is not understood, Nox1 may play a role in host defense. Nox1 expression is activated by bacterial lipopolysaccharide [10] and flagellins [11], and is more highly expressed in the distal colon, as expected from the bacterial colonization pattern [12].

Furthermore, Nox1 overactivity could be involved in IBD pathogenesis. Interleukin (IL)-10 deficient mice are a well-established model of spontaneously developing colitis, and experiments using these mice showed that the absence of IL-10 significantly facilitated Nox1 expression in association with increased interferon-γ expression before the development of spontaneous colitis [13]. In situ hybridization studies of colon biopsy samples obtained from patients with CD or UC revealed that Nox1 expression is localized in lesional lymphocytes [12]. Therefore, Nox1 may be involved in the onset of IBD, although the role of Nox1 in lymphocyte action and the possible link with IBD pathophysiology remains unknown.

Reactive Nitrogen Species in Inflammatory Bowel Disease

Mounting evidence indicates that the inflamed mucosa in IBD contains increased levels of RNSs like nitric oxide (NO). The increased serum and urinary nitrate levels in patients with IBD support the increase in NO production [14–16]. Increased NO production in the colon of patients with UC was confirmed by the detection of gas during colonoscopy [17]. Thus, high levels of NO metabolic end products such as nitrates and nitrites are found in the plasma, urine, and colonic lumen of IBD patients. However, diet and the presence of bacteria can contribute to the increase in NO levels; therefore, we should be aware that these indirect measurements might not accurately reflect the changes in NO levels in the intestinal mucosa.

NO is produced by a group of enzymes known as NO synthases (NOS), which consists of three isoforms, including neuronal NOS, endothelial NOS, and inducible NOS (iNOS). iNOS expression is markedly increased in the inflamed colonic mucosa of patients with IBD, suggesting that iNOS is a major source of increased NO production [18–20]. High iNOS activity was also reported in experimental models of colitis. Although it is controversial whether the effects of iNOS-induced NO are beneficial or detrimental in experimental colitis [21], the majority of studies using selective iNOS inhibitors have shown an improvement in experimental colitis. We reported that treatment with an iNOS-selective inhibitor, ONO-1714, ameliorated dextran sodium sulfate (DSS)-induced colitis in mice, suggesting that iNOS is involved in the progression of intestinal inflammation [22].

NO is inactivated by O_2^-, and the interaction of NO with O_2^- prevents O_2^--mediated hydroxylation reactions. Thus, the activity of NO may be dependent on the local O_2^- level and its scavenger systems [23]. Neutrophils can be induced to simultaneously produce NO and O_2^- in a concentrated localized manner by a variety of stimuli. NO and O_2^- then react to produce peroxynitrite, a potent and long-lived oxidant [24]. The peroxynitrite anion is cytotoxic because it inhibits mitochondrial electron transport, oxidizes protein sulfhydryl groups, initiates lipid peroxidation, and nitrates amino acids such as tyrosine, which affects many signal transduction pathways. The production of peroxynitrite can be indirectly inferred by the presence of nitrotyrosine (NT) residues [25]. NT was produced in the inflamed colonic mucosa of patients with IBD [19]. Keshavarzian et al. [26] demonstrated a correlation between the severity of colitis and NT levels, and reported that the actin cytoskeletal protein is a potential target of nitration. Ongoing studies in our lab include the analysis of oxidative-modified proteins such as NT-modified proteins. Our preliminary results identified NT-modified proteins in the inflamed colonic mucosa of 2,4,6-trinitrobenzene sulfonic acid-induced colitis, and was consistent with Keshavarzian et al. [26]. This indicates that identifying specific nitrated proteins would provide both mechanistic information regarding nitrosative stress and insight into downstream functional consequences.

Antioxidant Defense System in Inflammatory Bowel Disease

Oxidative stress is defined as an imbalance between the generation of ROS and antioxidant defense systems. In patients with IBD, depressed antioxidant defense systems appear to be involved in the development of intestinal inflammation.

Among the diverse endogenous mucosal antioxidant defense mechanisms, superoxide dismutases (SOD) are the primary defense against ROS. They convert O_2^- into hydrogen peroxide (H_2O_2), which is subsequently neutralized to oxygen and water by catalase or glutathione peroxidases (GPx). There are three SOD isoforms in humans: cytoplasmic copper/zinc (Cu/Zn)-SOD, mitochondrial manganese (Mn)-SOD, and extracellular (EC)-SOD. Accumulating evidence indicates that the protein levels and activities of Cu/Zn-SOD and EC-SOD are decreased in inflamed mucosa from IBD patients. On the other hand, it was reported that Mn-SOD is increased in inflamed mucosa of IBD patients compared with noninflamed mucosa and normal control mucosa. However, an increased Mn-SOD level is not always associated with increased resistance to oxidative stress because most of the increased Mn-SOD exists in an enzymatically inactive form [27]. These observations provide a rationale for SOD-based intervention therapy in IBD. In our previous study, treatment with Mn-SOD inhibited DSS-induced colonic mucosal injury in mice [28]. On the other hand, Cu/Zn-SOD-overexpressing mice showed significantly lower colonic neutrophilic myeloperoxidase activity than their nontransgenic littermates [29]. The administration of Cu/Zn-SOD suppressed the development of IBD-related colitis in experimental animal models [30, 31]. However, subsequent Cu/Zn-SOD clinical trials have been unsuccessful because of its low stability in plasma [32]. Lecithinized SOD (PC-SOD) has improved plasma stability, and is a potentially beneficial clinical treatment for IBD. The administration of PC-SOD achieves its ameliorative effect against DSS-induced colitis by decreasing the colonic level of ROS [33]. Moreover, the efficacy of PC-SOD for the treatment of UC patients has already been demonstrated by a phase II clinical study [34]. In the future, the clinical application of PC-SOD for IBD therapy is anticipated.

The glutathione system (glutathione, glutathione peroxidase, and glutathione reductase) is a key defense against H_2O_2 and other peroxides. In the presence of ROS, reduced GSH (GSH) is oxidized to glutathione disulphide (GSSG). This reaction is catalyzed by GPx and reversed by GSH reductase [35]. Reported changes in GSH levels vary among different studies. Iantomasi et al. [36] reported that GSH was decreased in the inflamed ileum of CD patients compared to noninflamed ileum and controls. Tsunada et al. [37] also demonstrated that GSH levels were significantly depleted in active colitis in UC patients as compared to the normal colon. On the other hand, some investigators have reported that the GSH level in inflamed mucosa samples from both UC and CD patients was unchanged compared to normal subjects [27, 37]. Sido et al. [38] identified increased GSSG levels in inflamed mucosa from patients with both UC and CD, consistent with previous reports [37, 39]. Holmes et al. [39] also reported decreased activity of γ-glutamylcysteine synthetase, the rate-limiting enzyme for glutathione biosynthesis, in inflamed mucosa from patients with IBD. Most studies revealed high levels of GPx in inflamed mucosa from both CD and UC patients, which appears to be a response against oxidative stress [27, 37, 39].

Thioredoxin (TRX) is one of the most important molecules that control the redox regulation system, and it contains a redox-active disulfide/dithiol (2 –SH groups that reside close to each other because of neighboring Cys residues) site: –Cys–Gly–Pro–Cys–. TRX plays a pivotal role in scavenging ROS with peroxiredoxins (Prxs) and prevents apoptosis of various cell types such as lymphocytes, monocytes, and epithelial cells. Tamaki et al. [40] reported that serum TRX levels were significantly higher in patients with both active UC and active CD than in normal controls, and TRX levels correlated with disease activity. They also revealed that the administration of TRX significantly ameliorated DSS-induced colitis and colonic inflammation in IL-10 deficient mice. Interestingly, the expression of TRX-interacting protein, which is a negative regulator of TRX, was significantly lower in the colonic mucosa of UC than in normal tissues [41]. These data indicate that TRX might be a new therapeutic molecule for IBD treatment.

Prxs are recently characterized members of the selenium (Se)-independent peroxidase superfamily that detoxify ROS and RNS. There are six described members of the Prx family in mammalian tissues, and these Prx proteins were further classified based on whether they contain one or two conserved Cys residues [42]. Our previous proteomic analysis of intestinal mucosa from mice with DSS-induced colitis demonstrated decreased expression of Prx-VI, the sole mammalian 1-Cys Prx [43]. Prx-VI is unlike the other mammalian members of the Prx family because it uses glutathione instead of TRX as a reductant. Our preliminary study showed decreased colonic expression of Prx-VI mRNA and protein in inflamed mucosa of IBD patients compared to normal colon tissues (fig. 1). The role of Prx-VI in the intestinal inflammation has not yet been described, but downregulation of Prx-VI might be involved in the pathogenesis of IBD.

Heme oxygenase-1 (HO-1) is one of three mammalian HO isozymes and is a stress-responsive protein induced by various oxidative agents, including oxidative stress, heat shock, ultraviolet radiation, ischemia-reperfusion, heavy metals, lipopolysaccharide, cytokines, NO, and heme (its substrate). The strong adaptive response of HO-1 to various stimuli suggests an entirely new paradigm in which HO-1 may play a significant role in protection against intestinal inflammation [44]. We showed that the expression of HO-1, which was mainly localized in mononuclear cells, was significantly increased in the colonic mucosa of patients with active UC as compared to normal mucosa [45]. In an animal model of colitis, HO activity and HO-1 expression

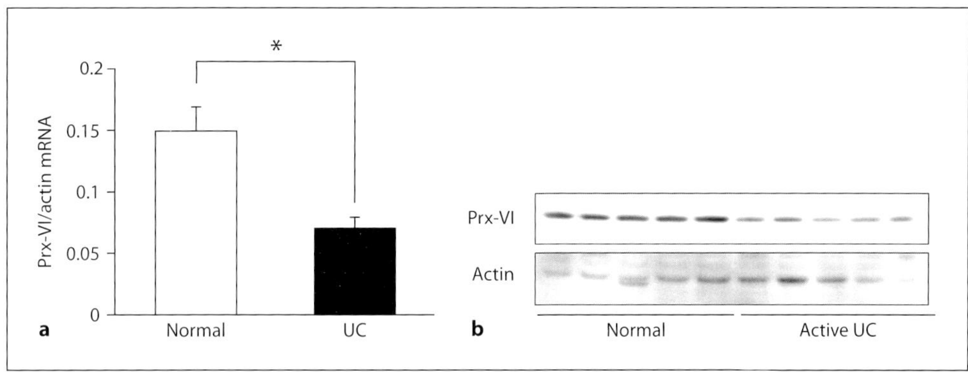

Fig. 1. **a** Expression of intestinal *prx-VI* mRNA determined by real-time polymerase chain reaction. To check *prx-VI* mRNA expression in inflamed mucosa, biopsy specimens were obtained from 12 patients in the active stage of UC and 13 normal subjects. *prx-VI* mRNA expression was normalized to the expression of *actin* mRNA. Data represent means ± SEM. * $p < 0.01$ vs. normal subjects. **b** Western blotting of Prx-VI. Colon tissue specimens were obtained during colonoscopy from patients with active UC and from normal subjects with colon cancer. Prx-VI protein levels in the colonic mucosa were determined by western blotting. Actin was used as the internal control.

markedly increased after induction of murine experimental colitis, and administration of an HO inhibitor potentiated colonic damage and inflammation [46–48]. A recent study showed that upregulation of HO-1 by cobalt protoporphyrin significantly reduced the intestinal injury induced by DSS [49]. These results indicate that HO-1 is an inducible protein responsible for host defense against intestinal inflammation and may be a novel therapeutic molecule for IBD.

Metallothionein (MT) is a Cys-rich low-molecular-weight protein that can act as a ROS scavenger. The role of MT in IBD has not yet been clarified. Although most investigations have demonstrated increased MT expression in patients with IBD [50, 51], other studies have yielded conflicting results regarding MT expression in IBD [52, 53]. Experimental colitis model studies using MT-deficient mice revealed that MT does not protect against the development of colitis [54, 55].

Conclusion

We have summarized the recent findings of the role of oxidative stress in intestinal inflammation in IBD. A growing body of evidence indicates that oxidants like free radicals play a key role in the pathophysiology of tissue injury in IBD, particularly in the initiation and perpetuation of inflammation and in subsequent tissue damage. Furthermore, we focused on the depressed antioxidant defense system involved in IBD pathogenesis. Regulating excessive free radical production and restoring the absent antioxidant defense system may be important therapeutic strategies for IBD treatment.

Conventional drugs for treating UC in humans such as sulfasalazine and 5-aminosalicylate (5-ASA) are potent ROS scavengers [56]. A recent study showed the ability of 5-ASA to induce HO-1 [57]. These agents produce anti-inflammatory effects through various functional actions, including the inhibition of NF-κB and inhibition of T cell proliferation; therefore, their therapeutic importance as ROS scavengers remains unclear. Interestingly, a recent study demonstrated that treating UC patients with 5-ASA compounds could prevent UC-associated colon carcinogenesis [58]. This suggested that at least one of these agents provided an excellent therapeutic effect by regulating oxidative stress. In the near future, we anticipate the development of novel IBD therapeutic strategies that are based on regulating oxidative stress.

References

1 Asakura K, Nishiwaki Y, Inoue N, Hibi T, Watanabe M, Takebayashi T: Prevalence of ulcerative colitis and Crohn's disease in Japan. J Gastroenterol 2009; 44:659–665.

2 Podolsky DK: Inflammatory bowel disease. N Engl J Med 2002;347:417–429.

3 Xavier RJ, Podolsky DK: Unravelling the pathogenesis of inflammatory bowel disease. Nature 2007; 448:427–434.

4 Naito Y, Takagi T, Yoshikawa T: Neutrophil-dependent oxidative stress in ulcerative colitis. J Clin Biochem Nutr 2007;41:18–26.

5 Naito Y, Takagi T, Yoshikawa T: Molecular fingerprints of neutrophil-dependent oxidative stress in inflammatory bowel disease. J Gastroenterol 2007; 42:787–798.

6 Karp SM, Koch TR: Oxidative stress and antioxidants in inflammatory bowel disease. Dis Mon 2006;52:199–207.

7 Rezaie A, Parker RD, Abdollahi M: Oxidative stress and pathogenesis of inflammatory bowel disease: an epiphenomenon or the cause? Dig Dis Sci 2007;52:2015–2021.

8 Kikuchi H, Hikage M, Miyashita H, Fukumoto M: NADPH oxidase subunit, gp91(phox) homologue, preferentially expressed in human colon epithelial cells. Gene 2000;254:237–243.

9 Suh YA, Arnold RS, Lassegue B, Shi J, Xu X, Sorescu D, Chung AB, Griendling KK, Lambeth JD: Cell transformation by the superoxide-generating oxidase Mox1. Nature 1999;401:79–82.

10 Kawahara T, Teshima S, Oka A, Sugiyama T, Kishi K, Rokutan K: Type I *Helicobacter pylori* lipopolysaccharide stimulates Toll-like receptor 4 and activates mitogen oxidase 1 in gastric pit cells. Infect Immun 2001;69:4382–4389.

11 Kawahara T, Kuwano Y, Teshima-Kondo S, Takeya R, Sumimoto H, Kishi K, Tsunawaki S, Hirayama T, Rokutan K: Role of nicotinamide adenine dinucleotide phosphate oxidase 1 in oxidative burst response to Toll-like receptor 5 signaling in large intestinal epithelial cells. J Immunol 2004;172:3051–3058.

12 Szanto I, Rubbia-Brandt L, Kiss P, Steger K, Banfi B, Kovari E, Herrmann F, Hadengue A, Krause KH: Expression of NOX1, a superoxide-generating NADPH oxidase, in colon cancer and inflammatory bowel disease. J Pathol 2005;207:164–176.

13 Kamizato M, Nishida K, Masuda K, Takeo K, Yamamoto Y, Kawai T, Teshima-Kondo S, Tanahashi T, Rokutan K: Interleukin 10 inhibits interferon gamma- and tumor necrosis factor alpha-stimulated activation of NADPH oxidase 1 in human colonic epithelial cells and the mouse colon. J Gastroenterol 2009;44:1172–1184.

14 Sasajima K, Yoshida Y, Yamakado S, Sato J, Miyashita M, Okawa K, Matsutani T, Onda M, Kawano E: Changes in urinary nitrate and nitrite during treatment of ulcerative colitis. Digestion 1996;57:170–173.

15 Oudkerk Pool M, Bouma G, Visser JJ, Kolkman JJ, Tran DD, Meuwissen SG, Pena AS: Serum nitrate levels in ulcerative colitis and Crohn's disease. Scand J Gastroenterol 1995;30:784–788.

16 Melichar B, Karlicek R, Tichy M: Increased urinary nitrate excretion in inflammatory bowel disease. Eur J Clin Chem Clin Biochem 1994;32:3–4.
17 Perner A, Nordgaard I, Matzen P, Rask-Madsen J: Colonic production of nitric oxide gas in ulcerative colitis, collagenous colitis and uninflamed bowel. Scand J Gastroenterol 2002;37:183–188.
18 Rachmilewitz D, Stamler JS, Bachwich D, Karmeli F, Ackerman Z, Podolsky DK: Enhanced colonic nitric oxide generation and nitric oxide synthase activity in ulcerative colitis and Crohn's disease. Gut 1995; 36:718–723.
19 Singer, II, Kawka DW, Scott S, Weidner JR, Mumford RA, Riehl TE, Stenson WF: Expression of inducible nitric oxide synthase and nitrotyrosine in colonic epithelium in inflammatory bowel disease. Gastroenterology 1996;111:871–885.
20 Kimura H, Hokari R, Miura S, Shigematsu T, Hirokawa M, Akiba Y, Kurose I, Higuchi H, Fujimori H, Tsuzuki Y, Serizawa H, Ishii H: Increased expression of an inducible isoform of nitric oxide synthase and the formation of peroxynitrite in colonic mucosa of patients with active ulcerative colitis. Gut 1998;42:180–187.
21 Cross RK, Wilson KT: Nitric oxide in inflammatory bowel disease. Inflamm Bowel Dis 2003;9:179–189.
22 Naito Y, Takagi T, Ishikawa T, Handa O, Matsumoto N, Yagi N, Matsuyama K, Yoshida N, Yoshikawa T: The inducible nitric oxide synthase inhibitor ONO-1714 blunts dextran sulfate sodium colitis in mice. Eur J Pharmacol 2001;412:91–99.
23 Moncada S, Palmer RM, Higgs EA: Nitric oxide: physiology, pathophysiology, and pharmacology. Pharmacol Rev 1991;43:109–142.
24 Beckman JS, Koppenol WH: Nitric oxide, superoxide, and peroxynitrite: the good, the bad, and ugly. Am J Physiol 1996;271:C1424–1437.
25 Ischiropoulos H: Biological tyrosine nitration: a pathophysiological function of nitric oxide and reactive oxygen species. Arch Biochem Biophys 1998;356:1–11.
26 Keshavarzian A, Banan A, Farhadi A, Komanduri S, Mutlu E, Zhang Y, Fields JZ: Increases in free radicals and cytoskeletal protein oxidation and nitration in the colon of patients with inflammatory bowel disease. Gut 2003;52:720–728.
27 Kruidenier L, Kuiper I, van Duijn W, Marklund SL, van Hogezand RA, Lamers CB, Verspaget HW: Differential mucosal expression of three superoxide dismutase isoforms in inflammatory bowel disease. J Pathol 2003;201:7–16.
28 Naito Y, Takagi T, Handa O, Ishikawa T, Matsumoto N, Yagi N: Role of superoxide and lipid peroxidation in the pathogenesis of dextran sulfate sodium-colitis in mice. ITE Lett 2001;2:663–667.
29 Kruidenier L, van Meeteren ME, Kuiper I, Jaarsma D, Lamers CB, Zijlstra FJ, Verspaget HW: Attenuated mild colonic inflammation and improved survival from severe DSS-colitis of transgenic Cu/Zn-SOD mice. Free Radic Biol Med 2003;34:753–765.
30 Segui J, Gironella M, Sans M, Granell S, Gil F, Gimeno M, Coronel P, Pique JM, Panes J: Superoxide dismutase ameliorates TNBS-induced colitis by reducing oxidative stress, adhesion molecule expression, and leukocyte recruitment into the inflamed intestine. J Leukoc Biol 2004;76:537–544.
31 Keshavarzian A, Morgan G, Sedghi S, Gordon JH, Doria M: Role of reactive oxygen metabolites in experimental colitis. Gut 1990;31:786–790.
32 Greenwald RA: Superoxide dismutase and catalase as therapeutic agents for human diseases. A critical review. Free Radic Biol Med 1990;8:201–209.
33 Ishihara T, Tanaka K, Tasaka Y, Namba T, Suzuki J, Ishihara T, Okamoto S, Hibi T, Takenaga M, Igarashi R, Sato K, Mizushima Y, Mizushima T: Therapeutic effect of lecithinized superoxide dismutase against colitis. J Pharmacol Exp Ther 2009;328:152–164.
34 Suzuki Y, Matsumoto T, Okamoto S, Hibi T: A lecithinized superoxide dismutase (PC-SOD) improves ulcerative colitis. Colorectal Dis 2008;10:931–934.
35 Meister A, Anderson ME: Glutathione. Annu Rev Biochem 1983;52:711–760.
36 Iantomasi T, Marraccini P, Favilli F, Vincenzini MT, Ferretti P, Tonelli F: Glutathione metabolism in Crohn's disease. Biochem Med Metab Biol 1994;53: 87–91.
37 Tsunada S, Iwakiri R, Ootani H, Aw TY, Fujimoto K: Redox imbalance in the colonic mucosa of ulcerative colitis. Scand J Gastroenterol 2003;38:1002–1003.
38 Sido B, Hack V, Hochlehnert A, Lipps H, Herfarth C, Droge W: Impairment of intestinal glutathione synthesis in patients with inflammatory bowel disease. Gut 1998;42:485–492.
39 Holmes EW, Yong SL, Eiznhamer D, Keshavarzian A: Glutathione content of colonic mucosa: evidence for oxidative damage in active ulcerative colitis. Dig Dis Sci 1998;43:1088–1095.
40 Tamaki H, Nakamura H, Nishio A, Nakase H, Ueno S, Uza N, Kido M, Inoue S, Mikami S, Asada M, Kiriya K, Kitamura H, Ohashi S, Fukui T, Kawasaki K, Matsuura M, Ishii Y, Okazaki K, Yodoi J, Chiba T: Human thioredoxin-1 ameliorates experimental murine colitis in association with suppressed macrophage inhibitory factor production. Gastroenterology 2006;131:1110–1121.
41 Takahashi Y, Masuda H, Ishii Y, Nishida Y, Kobayashi M, Asai S: Decreased expression of thioredoxin interacting protein mRNA in inflamed colonic mucosa in patients with ulcerative colitis. Oncol Rep 2007;18:531–535.

42 Rhee SG, Chae HZ, Kim K: Peroxiredoxins: a historical overview and speculative preview of novel mechanisms and emerging concepts in cell signaling. Free Radic Biol Med 2005;38:1543–1552.

43 Naito Y, Takagi T, Okada H, Omatsu T, Mizushima K, Handa O, Kokura S, Ichikawa H, Fujiwake H, Yoshikawa T: Identification of inflammation-related proteins in a murine colitis model by 2D fluorescence difference gel electrophoresis and mass spectrometry. J Gastroenterol Hepatol 2010;25(suppl): 144–148.

44 Naito Y, Takagi T, Yoshikawa T: Heme oxygenase-1: a new therapeutic target for inflammatory bowel disease. Aliment Pharmacol Ther 2004;20(suppl 1): 177–184.

45 Takagi T, Naito Y, Mizushima K, Nukigi Y, Okada H, Suzuki T, Hirata I, Omatsu T, Okayama T, Handa O, Kokura S, Ichikawa H, Yoshikawa T: Increased intestinal expression of heme oxygenase-1 and its localization in patients with ulcerative colitis. J Gastroenterol Hepatol 2008;23(suppl 2):S229–S233.

46 Naito Y, Takagi T, Tomatsuri N: Role of heme oxygenase-1 in dextran sulfate sodium-induced intestinal inflammation in mice. Gastroenterology 2003; 124(suppl):A–490.

47 Takagi T, Naito Y, Katada K: Heme oxygenase regulates the balance of inflammatory cytokines in dextran sulfate sodium-induced colitis. Gastroenterology 2004;126(suppl):A–564.

48 Wang WP, Guo X, Koo MW, Wong BC, Lam SK, Ye YN, Cho CH: Protective role of heme oxygenase-1 on trinitrobenzene sulfonic acid-induced colitis in rats. Am J Physiol Gastrointest Liver Physiol 2001; 281:G586–G594.

49 Berberat PO, YI AR, Yamashita K, Warny MM, Csizmadia E, Robson SC, Bach FH: Heme oxygenase-1-generated biliverdin ameliorates experimental murine colitis. Inflamm Bowel Dis 2005;11: 350–359.

50 Bruwer M, Schmid KW, Metz KA, Krieglstein CF, Senninger N, Schurmann G: Increased expression of metallothionein in inflammatory bowel disease. Inflamm Res 2001;50:289–293.

51 Dooley TP, Curto EV, Reddy SP, Davis RL, Lambert GW, Wilborn TW, Elson CO: Regulation of gene expression in inflammatory bowel disease and correlation with IBD drugs: screening by DNA microarrays. Inflamm Bowel Dis 2004;10:1–14.

52 Kruidenier L, Kuiper I, Van Duijn W, Mieremet-Ooms MA, van Hogezand RA, Lamers CB, Verspaget HW: Imbalanced secondary mucosal antioxidant response in inflammatory bowel disease. J Pathol 2003;201:17–27.

53 Ioachim E, Michael M, Katsanos C, Demou A, Tsianos EV: The immunohistochemical expression of metallothionein in inflammatory bowel disease. Correlation with HLA-DR antigen expression, lymphocyte subpopulations and proliferation-associated indices. Histol Histopathol 2003;18:75–82.

54 Tran CD, Ball JM, Sundar S, Coyle P, Howarth GS: The role of zinc and metallothionein in the dextran sulfate sodium-induced colitis mouse model. Dig Dis Sci 2007;52:2113–2121.

55 Oz HS, Chen T, de Villiers WJ, McClain CJ: Metallothionein overexpression does not protect against inflammatory bowel disease in a murine colitis model. Med Sci Monit 2005;11:BR69–BR73.

56 Craven PA, Pfanstiel J, Saito R, DeRubertis FR: Actions of sulfasalazine and 5-aminosalicylic acid as reactive oxygen scavengers in the suppression of bile acid-induced increases in colonic epithelial cell loss and proliferative activity. Gastroenterology 1987;92:1998–2008.

57 Horvath K, Varga C, Berko A, Posa A, Laszlo F, Whittle BJ: The involvement of heme oxygenase-1 activity in the therapeutic actions of 5-aminosalicylic acid in rat colitis. Eur J Pharmacol 2008;581:315–323.

58 Croog VJ, Ullman TA, Itzkowitz SH: Chemoprevention of colorectal cancer in ulcerative colitis. Int J Colorectal Dis 2003;18:392–400.

Tomohisa Takagi, MD, PhD
Molecular Gastroenterology and Hepatology
Graduate School of Medical Science
Kyoto Prefectural University of Medicine
465 Kajii-cho, Kawaramachi-Hirokoji, Kamigyo-ku
Kyoto 602-8566 (Japan)
Tel. +81 75 251 5508, Fax +81 75 251 0710, E-Mail takatomo@koto.kpu-m.ac.jp

Naito Y, Suematsu M, Yoshikawa T (eds): Free Radical Biology in Digestive Diseases.
Front Gastrointest Res. Basel, Karger, 2011, vol 29, pp 137–143

Free Radicals and Chronic Hepatitis C

Isao Sakaida

Department of Gastroenterology and Hepatology, Graduate School of Medicine, Yamaguchi University, Ube, Japan

Abstract

Transgenic mice transduced with the hepatitis C virus core gene develop hepatocellular carcinoma in the absence of inflammation in the liver. This phenomenon seems to be mainly related to oxidative stress in the liver, with mitochondria showing the highest reactive oxygen species (ROS) production in hepatocytes including hepatitis C virus core protein. Electron leakage is considered to occur from complex I toward the matrix and from complex III toward the cytoplasm, and impairment at these two sites is considered to be related to ROS production. Also, there is an increase in the calcium concentration in mitochondria that induces a decrease in the electric potential of the mitochondrial membranes and mitochondrial permeability transition, leading to an increase in ROS (superoxides) in mitochondria. Intrahepatic iron deposition is a well-known phenomenon in hepatitis C. Iron has important actions in the body but produces hydroxy radicals, a variation of ROS, by the Fenton reaction. Hydroxy radicals have been shown to be ROS with particularly strong histotoxicity, and iron is widely known to be a risk factor that exacerbates liver injury in chronic hepatitis and contributes to the development of hepatocellular carcinoma in the liver.

Production of Reactive Oxygen Species by Hepatitis C Virus Core Protein

In the first report of an in vivo enhancement of oxidative stress, Moriya et al. [1] documented an increase in lipid peroxides in the liver of transgenic mice transduced with the hepatitis C virus (HCV) core gene due to a decrease in reduced glutathione (GSH) and aging despite the absence of signs of inflammation in liver tissue. On the other hand, the expression of various antioxidant genes is enhanced, and this is considered to be a mechanism to avoid apoptosis under oxidative stress, important in relation to hepatocarcinogenesis.

Mitochondria are the sites showing the highest reactive oxygen species (ROS) production in hepatocytes as well as other cells. Usually, electrons are supplied to the electron transport system present in the mitochondrial inner membrane using pyruvic and succinic acid obtained from the glycolytic system as substrates. The electron transport system consists of 5 complexes (respiratory chain complexes), in which a hydrogen ion (H^+) is generated with each donation of an electron, and ATP is produced by the consequent electric gradient in the presence of ADP. In this process, the leakage of electrons is considered physiologically unavoidable, and if the electron transport system is impaired, electron transport is arrested, increasing the production of ROS. Electron leakage is considered to occur from complex I toward the matrix and from complex III toward the cytoplasm, and impairment at these two sites is considered to be related to ROS production.

ROS production was observed after the stimulation of Huh 7 cells, in which the expression of HCV core protein can be adjusted, with tertiary butyl hydroperoxide by confocal microscopy, and it was clarified that mitochondria are the primary site of ROS production due to the presence of HCV core protein, because ROS were stained more slowly than mitochondria [2]. Then, does HCV core protein actually damage electron transport respiratory chain complex I or III? Concerning this question, a decrease was shown in the activity of complex I in a liver mitochondrial fraction of transgenic mice expressing HCV structural proteins, and it was simultaneously reported that when glutamic and malic acids, which are substrates of complex I, were added to intact mitochondria, ROS production was significantly higher in transgenic mice and the GSH level in mitochondria was significantly lower [3]. This suggests that ROS are produced as HCV core protein damages mitochondrial complex I. However, the same paper also reported that HCV core protein was localized in the outer mitochondrial membrane. Schwer et al. [4] also reported that HCV core protein is localized in the mitochondria-associated membrane fraction, which is an interface between mitochondria and the endoplasmic reticulum (ER). This leads to a new question: how does HCV core protein in the outer mitochondrial membrane damage complex I in the inner mitochondrial membrane?

It was demonstrated that HCV core protein promotes calcium uptake by the calcium-selective ion channel called Ca^{2+} uniporter in the inner mitochondrial membrane using Huh 7 cells, which express HCV core protein as well as CYP2E1, and hepatic mitochondria from transgenic mice expressing HCV structural proteins. Interestingly, an enhancement of the Ca^{2+} uniporter activity was reported by another laboratory using cells in culture expressing HCV total protein [5]. These reports showed that an increase in the calcium concentration in mitochondria induces a decrease in the electric potential of the mitochondrial membranes and mitochondrial permeability transition, leading to an increase in ROS (superoxides) in mitochondria. Furthermore, nitrogen monoxide synthesis was shown to increase due to the influx of calcium into mitochondria, and this may impair oxidative phosphorylation within mitochondria. Many dehydrogenases are known to be activated with an increase in the calcium concentration in mitochondria, possibly promoting the supply of electrons to the electron transport system to enhance ROS production. It is important to note here that, because a Ca^{2+} uniporter inhibitor suppresses calcium influx into mitochondria and subsequent events, the influx of calcium via the Ca^{2+} uniporter initiates

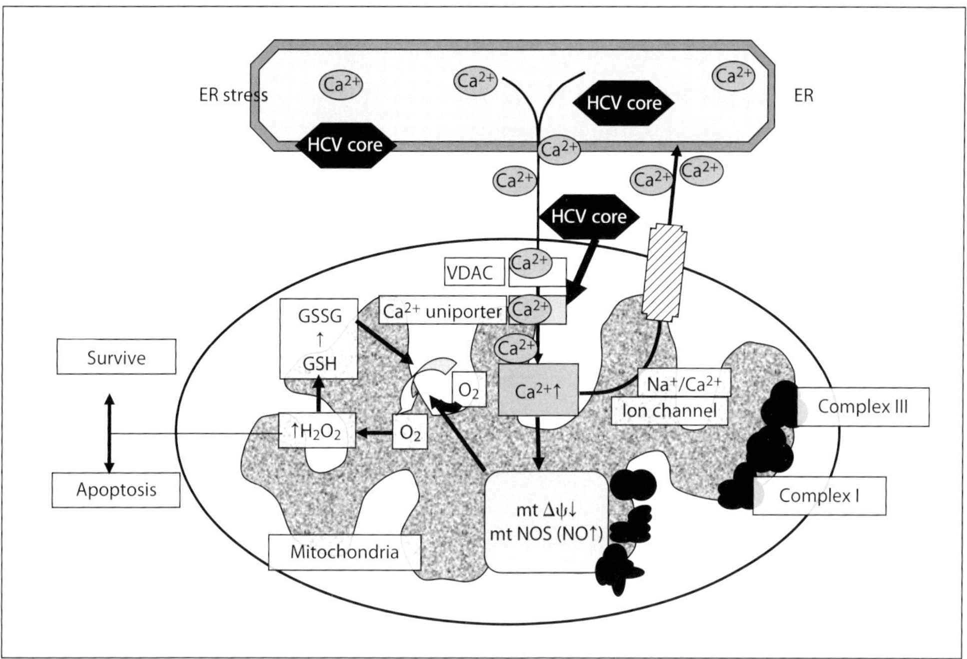

Fig. 1. ROS (H_2O_2) production in mitochondria with Ca^{2+} disturbance by HCV core protein in hepatocyte. mtΔψ = Electric potential of mitochondrial membranes; GSSG = glutathione disulphide.

mitochondrial injury, followed by oxidative stress. In other words, the suppression of calcium influx into mitochondria may be regarded as a molecular target for the control of oxidative stress induced by HCV protein. Since the localization of HCV protein in ER is considered to stress the ER and induce the disturbance of calcium regulation within it, the presence of core protein in the outer mitochondrial membrane adjacent to the ER appears reasonable. However, HCV protein is unlikely to directly affect the Ca^{2+} uniporter on the inner mitochondrial membrane. Also, calcium is considered to flow into mitochondria via the voltage-dependent anion channel (VDAC) present in the outer mitochondrial membrane, and the VDAC is considered to take up calcium into mitochondria by forming a complex with the Ca^{2+} uniporter. Therefore, HCV protein is speculated to promote calcium uptake by mitochondria as a result of activating the Ca^{2+} uniporter through some mechanism, but further research is necessary to clarify this point. These are summarized in figure 1.

Oxidative Stress-Related Factors in Hepatitis C

In the previous section, the mechanism of ROS production by HCV core protein was outlined, but the magnitude of oxidative stress shown by these experimental data is not very large. In fact, not all patients with chronic HCV infection uniformly develop hepatic fibrosis or cancer. Aging, alcohol, obesity, and iron accumulation are well-known factors exacerbating hepatitis C, but, interestingly, all these secondary factors augment oxidative stress, and the intensification of oxidative stress is considered necessary to modify the pathologic features of hepatitis C.

It has also been demonstrated by a large-scale cohort study that hepatic fibrosis is advanced regardless of the duration of infection in patients with type C chronic liver disease aged 65 years and above [6]. Mitochondrial injury and, particularly, reduced antioxidant activity are well-known age-associated phenomena as they are typically observed in Alzheimer's disease. It has also been shown in vitro that cells become more sensitive to ROS with aging [7]. Thus, aging is considered to increase oxidative stress and promote the progression of the disease in hepatitis C.

Cytochrome P450 (CYP, particularly CYP2E1), present in ER and microsomes, is a pathway of alcohol metabolism. Under alcohol loading, CYP2E1 produces ROS on receiving electrons from NADPH cytochrome P450 reductase. Therefore, in hepatitis C patients, chronic alcohol intake induces the expression of CYP2E1 and increases ROS production. Also, mitochondrial GSH is reduced in cells expressing HCV core protein+CYP2E1, suggesting that CYP2E1 induces mitochondria-dependent cell death by reducing the antioxidant activity in mitochondria.

Obesity can be classified into subcutaneous and visceral fat types, and the latter is particularly important for the development of fatty liver. Visceral fat type obesity induces insulin resistance and compensatory hyperinsulinemia, increases the influx of free fatty acids into the liver, and inhibits both fatty acid metabolism in the liver and the release of fatty acids from the liver, causing hepatic steatosis. In hepatitis C, also, a high body mass index and the occurrence of type 2 diabetes have been shown to be related to fat deposition and fibrosis of the liver [8]. Thus, hepatic steatosis due to obesity is considered an oxidative stress-exacerbating factor in hepatitis C.

Iron Metabolism in Hepatitis C

Iron absorbed from food in the upper small intestine is first transported to the liver via the portal vein. The liver not only stores iron, but also plays a crucial role in iron metabolism, such as in the production of iron carrier protein (Tf) and hepcidin known as a hormone that functions in the regulation of iron metabolism. Extracellular circulating iron in the plasma is present as soluble transferrin-bound iron, and when there is excess iron, as non-transferrin-bound iron, which is bound to serum proteins, such as albumin or citric acid. Hepatocytes express both transferrin receptor 1 (TfR1) and a homolog, TfR2. Under neutral pH conditions, Tf binds two iron atoms with high affinity, and circulates in the blood as Fe_2-Tf. Known routes for the import of Fe_2-Tf into hepatocytes are via TfR1 and TfR2, and a TfR-independent route.

Intrahepatic iron deposition is a well-known phenomenon in hepatitis C. Iron has important actions in the body but produces hydroxy radicals, a variation of ROS, by the Fenton reaction. Hydroxy radicals have been shown to be ROS with particularly strong histotoxicity, and iron is widely known to be a risk factor that exacerbates liver injury in chronic hepatitis. The mechanism of iron deposition with our transgenic mice using the total HCV genes (HCV TgM) is as follows [9].

HCV TgM showed significant increases in the intrahepatic and serum iron concentrations and a significant decrease in the intrasplenic iron concentration compared with control mice (C57BL/6) after the age of about 8 months. Moreover, iron deposition in liver tissue was more notable in hepatocytes than in Kupffer cells. No mechanism that specifically excretes iron has been clarified in vertebrates, and iron homeostasis is strictly controlled by iron absorption through the duodenal mucosal epithelium and iron release from the reticuloendothelial system. Associated with this, hepcidin, a peptide hormone generated and secreted by the liver, is known to regulate the iron content of the body by suppressing the expression of ferroportin, an iron transporter present on the vascular side of duodenal epithelial cells and in macrophages constituting the reticuloendothelial system, and controlling iron absorption via the duodenum and iron release from the reticuloendothelial system. The level of hepcidin mRNA expression in the liver was significantly lower in HCV TgM than in control mice, and, consistently, ferroportin expression in the duodenum and spleen was significantly higher in HCV TgM. Up to now, the transcription of hepcidin has been reported to be regulated by: (1) a system mediated by iron or bone morphogenic protein from hemojuvelin, a gene related to juvenile hemochromatosis, and SMAD signaling, and (2) a system mediated by the JAK-STAT signal from IL-6. HCV TgM did not develop inflammation in the liver but showed an increase in the expression of hepcidin mRNA when an inflammatory cytokine such as IL-6 was induced by the intraperitoneal administration of lipopolysaccharide, indicating that the second regulatory mechanism was not impaired. Also, the hepcidin promoter activity measured using hepatocytes from HCV TgM in the initial culture was significantly reduced compared with control mice. Moreover, the binding of CCAAT/enhancer binding protein (C/EBP), a transcription factor, to the promoter region of hepcidin was found to be reduced in HCV TgM. These phenomena are considered to have been due to an increase in the expression of C/EBP homology protein (CHOP) due to ROS because CHOP is known to form a dimer with C/EBP and inhibit the binding of C/EBP with its target DNA. The mechanism of iron excess in HCV TgM described above is summarized in figure 2.

Hepatocarcinogenesis due to Oxidative Stress

Oxidative stress due to hepatitis C is considered to contribute to a high rate of hepatocarcinogenesis. A state in which cells exhibiting oxidative DNA injury are not removed may be regarded as highly carcinogenic.

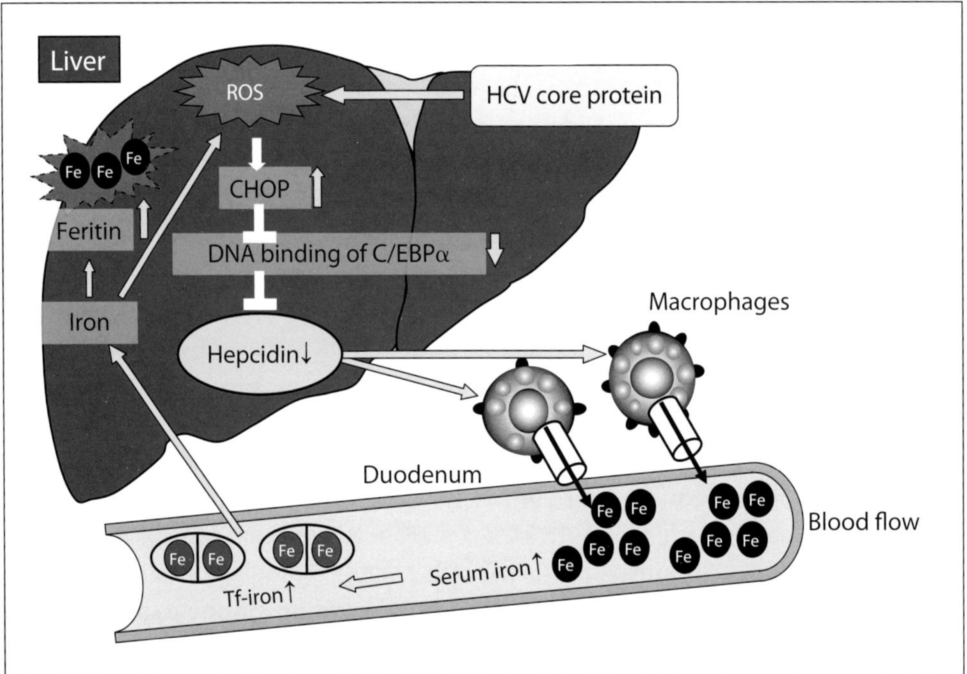

Fig. 2. Decreased hepcidin expression by ROS induces up regulation of FPN (ferroportion) in duodenum and macrophage resulting in increased iron absorption and iron release from macrophage. These phenomenon leads to excess accumulation of iron in HCV infected liver.

We evaluated how hepatocarcinogenesis occurs in a state under complex oxidative stress [10]. Four groups were established by feeding HCV TgM or control mice (C57BL/6) a normal (carbonyl iron content: 25 mg/kg) or iron overload (carbonyl iron content: 225 mg/kg) diet, and histological changes of the liver and the presence or absence of a liver tumor were evaluated after 12 months. The intrahepatic iron concentrations in HCV TgM after 12 months of iron overloading were similar to those in patients with chronic hepatitis C. After 12 months of iron overloading, HCV TgM showed increases in the intrahepatic deposits of 8-OHdG as well as lipid peroxides. It was also noted that the amount of intrahepatic 8-OHdG significantly increased with aging in not only iron-overloaded HCV TgM but also the other three groups. Eventually, liver tumors including hepatocelular carcinoma were noted in 45% (5/11) of the iron-overloaded HCV TgM alone. Thus, iron excess is induced by HCV itself and aging, and hepatocarcinogenesis is induced by the consequent exacerbation of oxidative stress and progression of oxidative DNA injury.

References

1 Moriya K, Fujie H, Shintani Y, Yotsuyanagi H, Tsutsumi T, Ishibashi K, Matsuura Y, Kimura S, Miyamura T, Koike K: The core protein of hepatitis C virus induces hepatocellular carcinoma in transgenic mice. Nat Med 1998;4:1065–1067.

2 Otani K, Korenaga M, Beard MR, Li K, Qian T, Showalter LA, Singh AK, Wang T, Weinman SA: Hepatitis C virus core protein, cytochrome P450 2E1, and alcohol produce combined mitochondrial injury and cytotoxicity in hepatoma cells. Gastroenterology 2005;128:96–107.

3 Korenaga M, Wang T, Li Y, Showalter LA, Chan T, Sun J, Weinman SA: Hepatitis C virus core protein inhibits mitochondrial electron transport and increases reactive oxygen species (ROS) production. J Biol Chem 2005;280:37481–37488.

4 Schwer B, Ren S, Pietschmann T, Kartenbeck J, Kaehlcke K, Bartenschlager R, Yen TS, Ott M: Targeting of hepatitis C virus core protein to mitochondria through a novel C-terminal localization motif. J Virol 2004;78:7958–7968.

5 Piccoli C, Scrima R, Quarato G, D'Aprile A, Ripoli M, Lecce L, Boffoli D, Moradpour D, Capitanio N: Hepatitis C virus protein expression causes calcium-mediated mitochondrial bioenergetic dysfunction and nitro-oxidative stress. Hepatology 2007;46:58–65.

6 Thabut D, Le Calvez S, Thibault V, Massard J, Munteanu M, Di Martino V, Ratziu V, Poynard T: Hepatitis C in 6,865 patients 65 years or older: a severe and neglected curable disease? Am J Gastroenterol 2006;101:1260–1267.

7 Tanguy S, Boucher F, Toufektsian MC, Besse S, de Leiris J: Aging exacerbates hydrogen peroxide-induced alteration of vascular reactivity in rats. Antioxid Redox Signal 2000;2:363–368.

8 Hourigan LF, Macdonald GA, Purdie D, Whitehall VH, Shorthouse C, Clouston A, Powell EE: Fibrosis in chronic hepatitis C correlates significantly with body mass index and steatosis. Hepatology 1999;29:1215–1219.

9 Nishina S, Hino K, Korenaga M, Vecchi C, Pietrangelo A, Mizukami Y, Furutani T, Sakai A, Okuda M, Hidaka I, Okita K, Sakaida I: Hepatitis C virus-induced reactive oxygen species raise hepatic iron level in mice by reducing hepcidin transcription. Gastroenterology 2008;134:226–238.

10 Furutani T, Hino K, Okuda M, Gondo T, Nishina S, Kitase A, Korenaga M, Xiao SY, Weinman SA, Lemon SM, Sakaida I, Okita K: Hepatic iron overload induces hepatocellular carcinoma in transgenic mice expressing the hepatitis C virus polyprotein. Gastroenterology 2006;130:2087–2098.

Isao Sakaida
Department of Gastroenterology and Hepatology
Graduate School of Medicine, Yamaguchi University
Ube (Japan)
Tel. +81 836 22 2238, Fax +81 836 22 2303, E-Mail sakaida@yamaguchi-u.ac.jp

Naito Y, Suematsu M, Yoshikawa T (eds): Free Radical Biology in Digestive Diseases.
Front Gastrointest Res. Basel, Karger, 2011, vol 29, pp 144–155

Free Radicals and Nonalcoholic Fatty Liver Disease/Nonalcoholic Steatohepatitis

Yoshio Sumida[a] · Yuji Naito[b] · Toshikazu Yoshikawa[b]

[a]Center for Digestive and Liver Diseases, Nara City Hospital, Nara, [b]Department of Molecular Gastroenterology and Hepatology, Kyoto Prefectural University of Medicine, Kyoto, Japan

Abstract

Nonalcoholic fatty liver disease (NAFLD) is one of the most common chronic liver diseases worldwide, and is now recognized as the hepatic manifestation of the metabolic syndrome. NAFLD represents a wide spectrum, ranging from simple steatosis, which is a benign condition, to steatohepatitis, which can progress to cirrhosis, hepatocellular carcinoma, and hepatic failure. Although the underlying pathophysiology of nonalcoholic steatohepatitis (NASH) remains to be clarified, NAFLD is characterized by two steps of liver injury: intrahepatic lipid accumulation (hepatic steatosis), and inflammatory progression to NASH (the 'two-hit' theory). Oxidative stress, which is determined by the balance between free radicals and antioxidants, is one of the most important causes. Sources of free radicals in NASH can include free fatty acids, inflammatory cytokines, adipokines, iron, myeloperoxidase, nitric oxide synthase, and so on. The measurement of systemic markers of oxidative stress in NAFLD/NASH has been attempted, but reliable, reproducible markers to reflect oxidative stress in the liver have not been realized. Although there are no established therapies for NASH, antioxidative therapies such as vitamin E and phlebotomy are promising. Recently, a phase 3, multicenter, randomized, placebo-controlled, double-blind clinical trial of pioglitazone, vitamin E or placebo for the treatment of adult NASH without diabetes proved that vitamin E was superior to placebo in the histological improvement. This article focuses on the role of free radicals in the diagnosis and therapy for NASH/NAFLD.

In 1980, Ludwig et al. [1] coined the term nonalcoholic steatohepatitis (NASH) to describe the morphologic pattern of liver injury in 20 patients evaluated at the Mayo Clinic. These patients had histologic evidence of alcoholic hepatitis on liver biopsy but no history of alcohol abuse. Nonalcoholic fatty liver disease (NAFLD) represents a wide spectrum of conditions ranging from simple steatosis (SS), which in general follows a benign nonprogressive clinical course, to steatohepatitis, which may progress to cirrhosis and end-stage liver disease (fig. 1). Now, NAFLD, the most common chronic liver disease in not only the Western world but also the Asia-Pacific region,

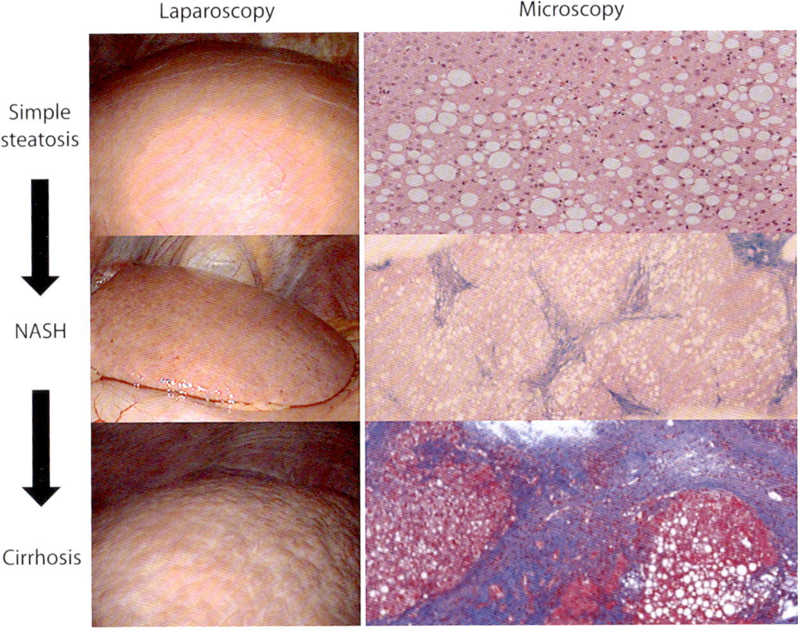

Fig 1. The broad spectrum of NAFLD. Three cases with SS, NASH, and cirrhosis, which were diagnosed by percutaneous liver biopsies under laparoscopy in Center for Digestive and Liver Diseases, Nara City Hospital.

is strongly associated with obesity, type 2 diabetes, and metabolic syndrome. Current best estimates make the prevalence of NAFLD approximately 20% and of NASH 2–3% in the general population.

In 1999, Matteoni et al. [2] divided NAFLD into four categories or types, based on the presence of steatosis, lobular inflammation, hepatocyte ballooning and Mallory-Denk bodies (MDB)/fibrosis; type 1: fatty liver alone, type 2: fat accumulation and lobular inflammation, type 3: fat accumulation and ballooning degeneration, and type 4: type 3 and either MDB or fibrosis. As originally reported, after a median follow-up period of 8.17 years, liver-related mortality of NAFLD (type 3 or 4) was 11 versus 1.7% in NAFLD (type 1 or 2). The current study, with a median follow-up period of 18.5 years, shows that liver-related mortality of NAFLD (type 3 or 4) increased to 17.5% in comparison with only 2.7% in NAFLD (type 1 or 2) (fig. 2) [3]. Thus, NAFLD (type 3 or 4) is now considered as a single group constituting NASH. These findings confirm that with longer follow-up periods, more NASH patients develop liver-related deaths, and that most patients with non-NASH are not subject to liver-related deaths. At the 45th Annual Meeting of Japan Society of Hepatology in June 2009, a consensus was reached that liver biopsy is the current gold standard for the diagnosis of NASH, and histopathological diagnosis of NASH is based on a combination of three features: (1) hepatic macrovesicular steatosis, (2) lobular inflammation, and (3) ballooning degeneration of hepatocytes.

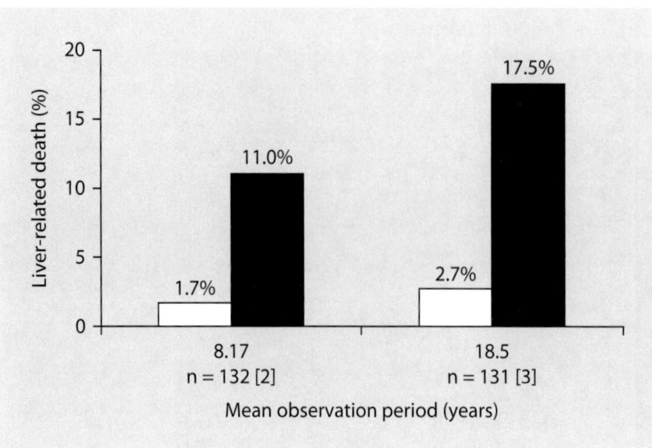

Fig 2. Liver-related mortality rate based on Matteoni's classification. Liver-related mortality was higher in patients with NASH (type 3 or 4; ■) than in those with non-NASH (type 1 or 2; □). Only NAFLD (type 3 or 4) should be considered as NASH.

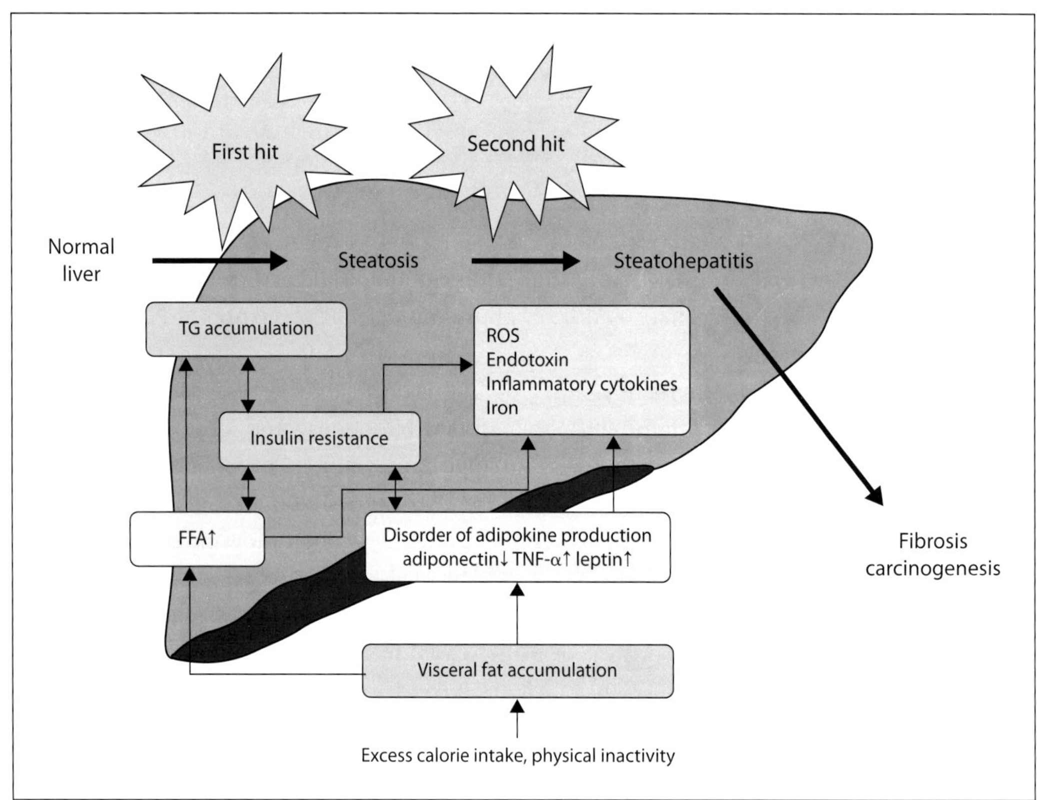

Fig 3. Pathogenesis of NASH/NAFLD (two hits theory). TG = Triglyceride. According to a two-hit hypothesis which best describes the progression from SS to NASH, fibrosis, or cirrhosis, increased intrahepatic triglyceride accumulation (due to increased synthesis, decreased export, or both) is followed by a second hit such as increased oxidative stress on hepatocytes, endotoxin, induction of proinflammatory cytokines, iron and so on.

Pathophysiology in Nonalcoholic Steatohepatitis

Pathophysiology of primary NASH still has not been completely clarified. According to the 'two hits' model of NASH pathogenesis proposed by Day and James [4], excessive triglyceride accumulation is the most likely first step. The second step may relate to an increase in oxidative stress, which, in turn, triggers liver cell necrosis and activation of hepatic stellate cells (HSCs), both leading to fibrosis and ultimately to the development of cirrhosis (fig. 3). Oxidative stress results from an imbalance between pro-oxidant and antioxidant chemical species that leads to oxidative damage of cellular macromolecules. The predominant pro-oxidant chemicals in fatty livers are singlet oxygen molecules, superoxide anions, hydrogen peroxide, and hydroxyl radicals: molecules collectively referred to as reactive oxygen species (ROS). ROS are relatively short-lived molecules that exert local effects. However, they can attack polyunsaturated fatty acids and initiate lipid peroxidation within the cell, which results in the formation of aldehyde by-products such as trans-4-hydroxy-2-nonenal (HNE) and malondialdehyde (MDA). When they bind to hepatocyte proteins, there is potentially harmful immune response from neoantigen formation by cross-linking cytokeratins to form MDB or activating HSCs to promote collagen synthesis and neutrophil chemotaxis stimulation.

Sources of Free Radicals in Nonalcoholic Steatohepatitis

The oxidation of free fatty acids (FFAs) is an important source of ROS in fatty livers [5, 6]. FFAs not incorporated into triglyceride are degraded by oxidation being catalyzed by enzymes located within three cellular compartments: mitochondria, peroxisomes, and microsomes [7]. Increased production of ROS in the presence of excess FFAs has been validated in not only animal models of NASH, but also in human livers with NASH. Transcription of enzymes catalyzing fatty acid β-oxidation in peroxisomes and mitochondria is regulated by the fatty acid-sensitive peroxisome proliferator activated receptor-α. Mitochondrial β-oxidation is the dominant oxidative pathway for the disposition of fatty acids under normal physiologic conditions, but can also be a major source of ROS. Several lines of evidence suggest that mitochondrial function is impaired in patients with NASH. Ultrastructural mitochondrial abnormalities have been documented in patients with NASH [8]. ROS are also produced when FFAs undergo ω-oxidation by cytochrome P450 enzyme within microsomes [9]. In addition, microsomal ω-oxidation of FFAs generates dicarboxylic acids uncoupling mitochondrial oxidative phosphorylation, thereby reducing the mitochondrial membrane potential. This decreases the efficiency of mitochondrial ATP production, enhancing vulnerability to other dangerous molecules capable of promoting depolarization of mitochondrial membranes, including tumor necrosis factor-α (TNF-α), together with other proapoptotic signals. Dicarboxylic acids are also peroxisome proliferator activated receptor-α ligands amplifying expression of fatty acid oxidizing enzymes, thus reinforcing expression of microsomal fatty acid oxidizing enzymes, such as cytochrome P450 2E1 (CYP2E1), and hence explaining why CYP2E1 expression and other microsomal enzymes are increased in NAFLD. The cumulative effect of extramitochondrial fatty acid oxidation is a further increase in oxidative stress and mitochondrial impairment.

A growing body of evidence supports a central role for inflammation and inflammatory cytokines in the development of NAFLD/NASH. The adipose tissue is an energy-storing organ that produces and secretes several bioactive substances known as adipokines, such as adiponectin, leptin, plasminogen activator inhibitor-1, and TNF-α. Adiponectin, a key molecule in obesity-related metabolic syndrome, has currently attracted much attention in the pathophysiology of metabolic syndrome because of its insulin-sensitizing function, anti-inflammatory properties, and anti-atherogenic function. Adiponectin also has a direct anti-inflammatory effect suppressing TNF-α production in the liver. Recent studies indicate that adiponectin can suppress oxidative stress, and that systemic oxidative stress, as measured by urinary 8-epi-prostaglandin F2α, correlates strongly with hypoadiponectinemia [10]. Leptin plays an important role in regulating the partitioning of fat between mitochondrial oxidation and triglyceride synthesis and also increases ROS production, especially by the respiratory chain. Leptin has a direct effect on HSC by triggering specific signal transduction systems of profibrotic genes such as collagen or tissue inhibitor of metalloprotease-1 altering collagen gene expression and increasing mitogenesis or has an indirect mechanism by acting on the Kupffer cells and sinusoidal endothelium releasing transforming growth factor-β_1 (TGF-β_1) stimulating fibrogenesis in activated HSCs. The mechanisms related to the cytokine-adipokine interaction in NASH are being intensely investigated. Insulin resistance is possibly regulated by proinflammatory cytokines, such as TNF-α and some adipokines, e.g. adiponectin and leptin [11]. Also, ROS mediates release of TNF-α by Kupffer cells, adipose tissue, and hepatocytes in NASH. TNF-α can also damage mitochondria and increase mitochondrial ROS formation. ROS-induced Fas ligand hepatocyte expression [12] and TNF-α-induced caspase activation contribute to apoptosis in NAFLD. TNF-α activates JNK and IKKβ, resulting in IRS serine phosphorylation and insulin resistance. Serum leptin and TNF-α levels were significantly higher, and adiponectin levels were significantly lower in patients with NASH than in control subjects.

It is conceivable that intestinally derived alcohol may contribute to fatty liver disease via generating ROS, because the intestinal production of ethanol by small intestinal bacterial overgrowth is increased in obesity [13]. Plasma myeloperoxidase (MPO), a neutrophil enzyme that generates aggressive ROS, was elevated in NASH patients. The accumulation of MPO-mediated oxidation products induced various chemokines and hepatic neutrophil infiltration, thus contributing to the development of NASH [14].

Altered iron parameters (serum ferritin levels or transferring saturation) are frequently detected in patients with NAFLD. It has been hypothesized that iron induces oxidative stress by catalyzing production of ROS through Fenton reaction, and increased hepatic iron deposition could be a part of the pathogenesis of NAFLD [15]. A recent study from Italy suggests that predominantly hepatocellular iron deposition is associated with more severe liver damage in NAFLD [16]. Although the precise mechanisms underlying hepatic iron deposition in NASH/NAFLD remain unknown, several factors including genetic factors (HFE mutation), insulin resistance, disturbed retinoid signals [17], dysregulation of hepcidin synthesis [18], copper deficiency, and erythrophagocytosis by Kupffer cells, may be responsible for hepatic iron accumulation in NASH [15].

Indicators of Oxidative Stress

Several groups have attempted to elucidate whether measurement of systemic markers of oxidative stress may reflect the levels of oxidative stress present in the liver (table 1). Most groups have used either the method to measure systemic levels of stable lipid byproducts of ROS activity such as lipid peroxides and thiobarbituric acid-reacting substance (TBARS) or 'total antioxidant status' with mixed results (table 1). Several groups reported higher levels of lipid peroxidation products in NASH patients compared with SS patients or healthy people. Hepatic expression of 8-hydroxydeoxyguanosine (8-OHdG), as reliable markers of oxidative DNA damage, was increased in NAFLD patients [19], especially in NASH patients [20]. Fujita et al. [20] suggested that hepatic iron overload and insulin resistance were independently associated with elevated 8-OHdG. On the other hand, antioxidative properties such as superoxide dismutase activity, catalase, and glutathione seem to be decreased in serum or red blood cells. As we previously reported, thioredoxin (TRX), an oxidative stress-inducible thiol-containing protein, which has important roles in the redox regulation, is also significantly elevated in the NASH patients' serum, compared to those with SS or healthy people [21]. Although the existence of oxidative stress in the liver of NASH patients has been clearly demonstrated, future studies will need to characterize which markers are better suited for NASH diagnosis.

Antioxidative Therapies for Nonalcoholic Steatohepatitis

There has been no firm evidence-based treatment for NASH/NAFLD. The depletion of antioxidants within hepatocytes resulting in impaired ROS inactivation is the basis for antioxidant supplementation as a potential treatment for NASH. Several encouraging pilot studies of various agents indicate potential beneficial effects which may

Table 1. Indicators of oxidative stress in patients with NAFLD/NASH

First author and year	Patients	Sample	Parameters	NASH vs. control	NASH vs. SS
Loguercio 2001	NAFLD (n=81) NASH/SS=34:14	serum RBC	HNE, MDA	↑	↑
Seki 2002	NASH (n=17) SS (n=23) control (n=7)	liver	8-OHdG	↑↑	→
			HNE	↑	→~↑
Sumida 2003	NASH (n=25) SS (n=15) control (n=17)	serum	TRX	↑	↑
Koruk 2004	NASH (n=18) control (n=16)	serum	GSH, MDA, NO	↑	
			SOD	↓	
			GSH-Px, GR	→	
Chalassani 2004	NASH (n=21) control (n=19)	serum	oxidized LDL, TBARS	↑	
Horoz 2005	NASH (n=22) control (n=22)	serum	total antioxidant response (TAR)	↓	
			total plasma peroxide	↑	
Yesilova 2005	NAFLD (n=51) control (n=30) viral hepatitis (n=30)	serum	MDA	↑	
			Cu,Zn- SOD, catalase, coenzyme Q	↓	
Machado 2008	NASH (n=43) control (n=33)	serum RBC	8-OHdG	→	
			reduced GSH, reduced GSH/ oxidized GSH ratio	↓	
			total antioxidant status (TAS), vitamin E	↑	
			GSH-Px, GR	→	
Fujita 2009	NASH (n=38) SS (n=24) control (n=10)	liver	8-OHdG	↑	↑

SS= Simple steatosis; HNE= 4-hydroxy-2-nonenal; MDA= malonedialdehyde; 8-OHdG= 8-hydroxy deoxyguanosine; TRX= thioredoxin; GSH= glutathione; NO= nitric oxide; TBARS= thiobarbituric acid-reactive substances; SOD= superoxide dismutase; GSH-Px= glutathione peroxidase; GR= glutathione reductase.

be related to their antioxidant effects. These include vitamin E, N-acetyl cysteine, probucol, betaine, and iron depletion through phlebotomy.

The lipid-soluble antioxidant, α-tocopherol (vitamin E), has been shown to inhibit lipid peroxidation and suppress inflammatory cytokines such as TNF-α, and its use in the treatment of NASH/NAFLD has been studied (table 2). The first human trial of vitamin E in NAFLD was in a pediatric group of patients [22]. In this trial, vitamin E resulted in normalization of alanine aminotransferase (ALT) levels in 11 obese children. However, a randomized controlled trial (RCT) of vitamin E combined with lifestyle modification in 28 pediatric patients with NAFLD found no overall improvement in ALT levels compared with placebo [23]. In adults, a group of biopsy-proven NASH patients were randomized to vitamin E (1,000 IU/day) plus vitamin C (1,000 mg/day) or placebo for 6 months. The patients showed no improvement in serum aminotransferase levels but repeat liver biopsies at the end of the trial demonstrated decreased fibrosis within the vitamin group (especially in patients with diabetes) [24]. In Japan, a study with vitamin E administration (300 mg/day during 1 year) confirmed improvement in liver tests with reduction of plasma levels of TGF-β_1, but revealed no change in degree of steatosis, fibrosis and inflammation on follow-up liver biopsy [25]. Kawanaka et al. [26] revealed that serum transaminase activities, TRX, and TBARS were significantly decreased after the 6-month course of vitamin E therapy. Recently, the Nonalcoholic Steatohepatitis Clinical Research Network conducted the Pioglitazone versus Vitamin E versus Placebo for the Treatment of Nondiabetic Patients with Nonalcoholic Steatohepatitis trial, a phase 3, multicenter, randomized, placebo-controlled, double-blind clinical trial of pioglitazone or vitamin E for the treatment of adults without diabetes who had biopsy-confirmed NASH. By histological analysis, vitamin E was superior to placebo in regard to the resolution of NASH, and there was no benefit of pioglitazone over placebo [27]. In the future, we should examine whether vitamin E treatment may prevent hepatic carcinogenesis or hepatic failure, leading to improved overall survival rate in NAFLD patients.

Phlebotomy targeting hepatic iron overload with resultant increased oxidative stress is another potential therapeutic modality that has been investigated in the treatment of NASH. In a few pilot studies in NASH patients including ours [28], quantitative phlebotomies were performed with induction of iron depletion and reduction in serum ferritin levels. Results of these studies suggest that iron reduction therapy by phlebotomy can lead to improvement in aminotransferase levels and insulin resistance; unfortunately, repeat liver biopsies were not performed [15].

Betaine, a nutritional supplement that is a component of the metabolic cycle of methionine, raises S-adenosylmethionine levels, which may reduce hepatic steatosis. An 8-week trial of betaine supplementation in patients with NASH compared with placebo showed reductions in ALT, AST, and GGT levels, and steatosis in the treatment group [29]. Another placebo-controlled study included 191 patients

Table 2. Vitamin E for the treatment of NAFLD/NASH

First author and year	Study design	Patient	Age, years	Therapy	Dose, /day
Lavine 2000	pilot study	NAFLD (n = 11)	12.4±1.6	vit. E	400–1,200 IU
Hasegawa 2001	pilot study	NASH (n = 12)	40±4	vit. E	300 mg
		NAFL (n = 10)	38±3		
Harrison 2003	RCT	NASH (n = 45)	52.5 50.2	vit. E+C (n = 23)	vit. E: 1,000 IU vit. C: 1,000 mg
				placebo (n = 22)	
Kugelmas 2003	pilot study	NASH (n = 16)	41.5±2.6 54.8±2.3	diet+vit. E (n = 9)	800 IU
				diet (n = 7)	
Vajro 2004	RCT	NAFLD (n = 28)	10.7±3.45	diet+vit.E (n = 14)	400 mg (2 months) 100 mg (3 months)
			9.88±3.97	diet+placebo (n = 14)	
Kawanaka 2004	pilot study	NASH (n = 10)	60.6±13.3	vit. E	300 mg
Sanyal 2010	RCT	NASH (n = 247)	46.6±12.1	vit. E (n = 84)	800 IU
			47.0±12.6	PIO (n = 80)	30 mg
			45.4±11.2	placebo (n = 83)	

NA = Not assessed; HA = hyaluronic acid; PIO = pioglitazone. Values for age and ALT are expressed as mean ± SD.

Treatment duration	ALT, IU/l		p	Histologic features	Other findings
	before	after			
4–10 months	175±106	40±26	< 0.001	NA	
1 years	161±14	37±4	< 0.01		
	43±3	41±3	NS	steatosis improved in 6 of 9 NASH patients; inflammation and fibrosis improved in 5 of them	TGF-β_1↓
6 months	92.3	no decrease		fibrosis significantly improved (especially diabetic patients)	
	109	significant decrease		no significant change in inflammation and fibrosis	
12 weeks	64.78±9.54	significant decrease in all patients no difference between two groups		NA	HA↓ IL-6↓
	56.57±7.39				
5 months	72.58±17.46	31.90% ↓ (at 2 months) 24.09% ↓ (at 5 months)		NA	
	76.00±23.45	26.36% ↓ (at 2 months) 29.92% ↓ (at 5 months)			
		no difference between two groups			
6 months	98.3±22.7	49.4±23.3	<0.001	NA	TRX ↓ TBARS↓
96 weeks	86±52	−37.0	0.001 vs. placebo	43% improved (p = 0.001 vs. placebo)	
	82±45	−40.8	<0.0001 vs. placebo	34% improved (p = 0.04 vs. placebo)	
	83±49	−20.1		19% improved	

with NAFLD treated for 8 weeks with combination of betaine glucuronate (300 mg/day), diethanolamine and nicotinamide ascorbate. Significant improvement was noted in liver tests and degree of steatosis in the treated group. Probucol, which is a lipid-lowering agent with strong antioxidant properties, appears to be efficacious in decreasing ALT levels in NASH patients, according to a double-blind randomized controlled study by Merat et al. [30]. This group also revealed that hepatic inflammation was reduced in NASH patients after use of probucol for 1 year. N-acetyl cysteine, a precursor of glutathione, ameliorates NASH histology in animal models of NASH. There have been few studies to examine the efficacy in human NASH.

Overall, antioxidants represent a novel class of medications that have shown some promising initial results in the treatment of NASH, but further well-designed large RCTs are required, and it is likely that these therapies will be used in combination regimens with other agents that affect insulin sensitivity. The research agenda for the future includes establishing the role of free radicals in NASH/NAFLD, identifying better predictors of the disease, and defining effective treatment.

References

1 Ludwig J, Viggiano TR, McGill DB, Oh BJ: Nonalcoholic steatohepatitis: Mayo Clinic experiences with a hitherto unnamed disease. Mayo Clin Proc 1980;55:434–438.

2 Matteoni CA, Younossi ZM, Gramlich T, Boparai N, Liu YC, McCullough AJ: Nonalcoholic fatty liver disease: a spectrum of clinical and pathological severity. Gastroenterology 1999;116:1413–1419.

3 Rafiq N, Bai C, Fang Y, Srishord M, McCullough A, Gramlich T, Younossi ZM: Long-term follow-up of patients with nonalcoholic fatty liver. Clin Gastroenterol Hepatol 2009;7:234–238.

4 Day CP, James OF: Steatohepatitis: a tale of two 'hits'? Gastroenterology 1998;114:842–845.

5 Donnelly KL, Smith CI, Schwarzenberg SJ, Jessurun J, Boldt MD, Parks EJ: Sources of fatty acids stored in liver and secreted via lipoproteins in patients with nonalcoholic fatty liver disease. J Clin Invest 2005;115:1343–1351.

6 Shiota G, Tsuchiya H: Pathophysiology of NASH: insulin resistance, free fatty acids and oxidative stress. J Clin Biochem Nutr 2006;38:127–132.

7 Browning JD, Horton JD: Molecular mediators of hepatic steatosis and liver injury. J Clin Invest 2004;114:147–152.

8 Caldwell SH, de Freitas LA, Park SH, Moreno ML, Redick JA, Davis CA, Sisson BJ, Patrie JT, Cotrim H, Argo CK, Al-Osaimi A: Intramitochondrial crystalline inclusions in nonalcoholic steatohepatitis. Hepatology 2009;49:1888–1895.

9 Chalasani N, Gorski JC, Asghar MS, Asghar A, Foresman B, Hall SD, Crabb DW: Hepatic cytochrome P450 2E1 activity in nondiabetic patients with nonalcoholic steatohepatitis. Hepatology 2003;37:544–550.

10 Kamada Y, Takehara T, Hayashi N: Adipocytokines and liver disease. J Gastroenterol 2008;43:811–822.

11 Watanabe S, Yaginuma R, Ikejima K, Miyazaki A: Liver diseases and metabolic syndrome. J Gastroenterol 2008;43:509–518.

12 Feldstein AE, Canbay A, Angulo P, Taniai M, Burgart LJ, Lindor KD, Gores GJ: Hepatocyte apoptosis and fas expression are prominent features of human nonalcoholic steatohepatitis. Gastroenterology 2003;125:437–443.

13 Nair S, Cope K, Risby TH, Diehl AM: Obesity and female gender increase breath ethanol concentration: potential implications for the pathogenesis of nonalcoholic steatohepatitis. Am J Gastroenterol 2001;96:1200–1204.

14 Rensen SS, Slaats Y, Nijhuis J, Jans A, Bieghs V, Driessen A, Malle E, Greve JW, Buurman WA: Increased hepatic myeloperoxidase activity in obese subjects with nonalcoholic steatohepatitis. Am J Pathol 2009;175:1473–1482.

15 Sumida Y, Yoshikawa T, Okanoue T: Role of hepatic iron in non-alcoholic steatohepatitis. Hepatol Res 2009;39:213–222.

16 Valenti L, Fracanzani AL, Bugianesi E, Dongiovanni P, Galmozzi E, Vanni E, Canavesi E, Lattuada E, Roviaro G, Marchesini G, Fargion S: HFE genotype, parenchymal iron accumulation, and liver fibrosis in patients with nonalcoholic fatty liver disease. Gastroenterology 2010;138:905–912.

17 Tsuchiya H, Akechi Y, Ikeda R, Nishio R, Sakabe T, Terabayashi K, Matsumi Y, Ashla AA, Hoshikawa Y, Kurimasa A, Suzuki T, Ishibashi N, Yanagida S, Shiota G: Suppressive effects of retinoids on iron-induced oxidative stress in the liver. Gastroenterology 2009;136:341–350.

18 Mitsuyoshi H, Yasui K, Harano Y, Endo M, Tsuji K, Minami M, Itoh Y, Okanoue T, Yoshikawa T: Analysis of hepatic genes involved in the metabolism of fatty acids and iron in nonalcoholic fatty liver disease. Hepatol Res 2009;39:366–373.

19 Seki S, Kitada T, Yamada T, Sakaguchi H, Nakatani K, Wakasa K: In situ detection of lipid peroxidation and oxidative DNA damage in non-alcoholic fatty liver diseases. J Hepatol 2002;37:56–62.

20 Fujita N, Miyachi H, Tanaka H, Takeo M, Nakagawa N, Kobayashi Y, Iwasa M, Watanabe S, Takei Y: Iron overload is associated with hepatic oxidative damage to DNA in nonalcoholic steatohepatitis. Cancer Epidemiol Biomarkers Prev 2009;18:424–432.

21 Sumida Y, Nakashima T, Yoh T, Furutani M, Hirohama A, Kakisaka Y, Nakajima Y, Ishikawa H, Mitsuyoshi H, Okanoue T, Kashima K, Nakamura H, Yodoi J: Serum thioredoxin levels as a predictor of steatohepatitis in patients with nonalcoholic fatty liver disease. J Hepatol 2003;38:32–38.

22 Lavine JE: Vitamin E treatment of nonalcoholic steatohepatitis in children: a pilot study. J Pediatr 2000; 136:734–738.

23 Vajro P, Mandato C, Franzese A, Ciccimarra E, Lucariello S, Savoia M, Capuano G, Migliaro F: Vitamin E treatment in pediatric obesity-related liver disease: a randomized study. J Pediatr Gastroenterol Nutr 2004;38:48–55.

24 Harrison SA, Torgerson S, Hayashi P, Ward J, Schenker S: Vitamin E and vitamin C treatment improves fibrosis in patients with nonalcoholic steatohepatitis. Am J Gastroenterol 2003;98:2485–2490.

25 Hasegawa T, Yoneda M, Nakamura K, Makino I, Terano A: Plasma transforming growth factor-beta1 level and efficacy of alpha-tocopherol in patients with non-alcoholic steatohepatitis: a pilot study. Aliment Pharmacol Ther 2001;15:1667–1672.

26 Kawanaka M, Mahmood S, Niiyama G, Izumi A, Kamei A, Ikeda H, Suehiro M, Togawa K, Sasagawa T, Okita M, Nakamura H, Yodoi J, Yamada G: Control of oxidative stress and reduction in biochemical markers by Vitamin E treatment in patients with nonalcoholic steatohepatitis: a pilot study. Hepatol Res 2004;29:39–41.

27 Sanyal AJ, Chalasani N, Kowdley KV, McCullough A, Diehl AM, Bass NM, Neuschwander-Tetri BA, Lavine JE, Tonascia J, Unalp A, Van Natta M, Clark J, Brunt EM, Kleiner DE, Hoofnagle JH, Robuck PR; NASH CRN: Pioglitazone, vitamin E, or placebo for nonalcoholic steatohepatitis. New Engl J Med 2010; 362:1675–1685.

28 Sumida Y, Kanemasa K, Fukumoto K, Yoshida N, Sakai K, Nakashima T, Okanoue T: Effect of iron reduction by phlebotomy in Japanese patients with nonalcoholic steatohepatitis: a pilot study. Hepatol Res 2006;36:315–321.

29 Abdelmalek MF, Angulo P, Jorgensen RA, Sylvestre PB, Lindor KD: Betaine, a promising new agent for patients with nonalcoholic steatohepatitis: results of a pilot study. Am J Gastroenterol 2001;96:2711–2717.

30 Merat S, Aduli M, Kazemi R, Sotoudeh M, Sedighi N, Sohrabi M, Malekzadeh R: Liver histology changes in nonalcoholic steatohepatitis after one year of treatment with probucol. Dig Dis Sci 2008; 53:2246–2250.

Yoshio Sumida
Center for Digestive and Liver Diseases, Nara City Hospital
Higashi Kidera-cho 1-50-1
Nara 630-8305 (Japan)
Tel. +81 742 24 1251, Fax +81 742 22 2478, E-Mail sumida@nara-jadecom.jp

Naito Y, Suematsu M, Yoshikawa T (eds): Free Radical Biology in Digestive Diseases.
Front Gastrointest Res. Basel, Karger, 2011, vol 29, pp 156–163

Role of Free Radicals in Pancreatitis

Shin Hamada · Tooru Shimosegawa

Division of Gastroenterology, Tohoku University Graduate School of Medicine, Sendai, Japan

Abstract

Acute pancreatitis and chronic pancreatitis show distinct clinical features, but they share common pathological changes in the pancreas. The establishment of inflammation within the pancreas involves a multiple process, and free radicals play crucial roles during the exacerbation of inflammation. Reactive oxygen species and reactive nitrogen species are produced by acinar cells, pancreatic stellate cells and infiltrating inflammatory cells. These molecules further activate downstream signaling pathway. In the case of acute pancreatitis, acinar cell damage leads to free radical production, which triggers cytokine production. Infiltrating neutrophils further create free radicals by 'respiratory burst', resulting in amplification of tissue damage. Excessive free radicals also contribute to multiple organ failure. In case of chronic pancreatitis, free radicals help to establish the persistent inflammation referred to as 'necrosis-fibrosis sequences'. Pancreatic stellate cells are activated by free radicals, leading to the increased production of cytokines and extracellular matrix, finally developing irreversible fibrosis in the pancreas. Free radicals exert these effects not only by direct injury to the cellular components, but also by activating multiple signaling pathways, such as NF-κB or mitogen-activated protein kinase. Inhibition of free radicals by free radical scavengers or inhibitors of reactive oxygen species-generating enzyme shows therapeutic effects in the experimental model, and clinical application is also expected. Targeting the free radical-generating system would be a novel therapeutic strategy against acute and chronic pancreatitis.

Inflammatory diseases of the pancreas have two major sides. One is acute pancreatitis, which manifests acute symptoms derived from rapidly spreading inflammatory changes in pancreas. Severe acute pancreatitis is a life-threatening situation, which leads to multiple organ failure, as a result of catastrophic positive feedback loop of inflammation [1]. The other, chronic pancreatitis, is defined as a progressive, irreversible destruction of exocrine and endocrine pancreas tissues by repeated inflammation, so-called 'necrosis-fibrosis sequence' [2]. Loss of pancreas function results in malabsorption of nutrients and pancreatic diabetes [3, 4], which impairs patients' quality of life, and furthermore, raises risk of pancreatic cancer [5].

These two sides of pancreatic inflammatory diseases have distinct clinical characteristics, but in both cases the existence of inflammation is a common basis of pathophysiology. In the early pancreatitis responses, activation of zymogens is critical for acinar cells [6], and once the defensive mechanism has been overwhelmed by continuous activation, consequent destruction of acinar cells and recruitment of inflammatory cells establish pancreatitis. Free radicals produced within the pancreas contribute to this process. Free radicals include reactive oxygen species (ROS) and reactive nitrogen species (RNS). ROS consist of superoxide, singlet oxygen, hydrogen peroxide and so on. RNS consist of nitric oxide (NO) and other nitrates. These molecules are known to activate multiple signaling pathways and have a direct effect on biomolecules such as protein, plasma membrane and DNA, which are essential for the proper cellular activity. Free radicals are capable of inducing tissue injury in multiple organs including the pancreas [7], which contributes to the exacerbation of acute and chronic inflammation. This chapter handles the pivotal role of free radicals in acute and chronic pancreatitis as a promoting factor of the inflammatory process.

Acute Pancreatitis and Free Radicals

Since free radicals are highly reactive and directly assault the cellular component, the contribution of free radicals to the acute inflammatory process has been extensively studied. According to a previous report, upon the induction of experimental acute pancreatitis in common pancreaticobiliary duct ligation model by using the rat, the concentration of malondialdehyde, a lipid peroxidation molecule, in the pancreas was elevated and the concentration of glutathione, a free radical scavenger, was reduced [8], which represents increased production of free radicals within pancreas. Transarterial administration of xanthine oxidase (XOD) and hypoxanthine via the celiac artery in the rat, which gives direct oxidative stress to the pancreas, resulted in no remarkable histological changes compatible with acute pancreatitis [9]. Based on these observations, oxidative stress is considered to act as a promoter of pancreatic injury, not as a trigger of initial inflammation.

The mechanisms of free radical induction after the onset of acute pancreatitis have been explored, and several pathways were uncovered. Stimulation of the rat pancreatic acinar-derived AR42J cells by cerulein, the cholecystokinin analog, resulted in the ROS production via the nicotinamide adenine dinucleotide phosphate (NADPH) oxidase induction, which is one of the major ROS-generating enzymes [10]. This in vitro experiment revealed that the acinar cell itself could be a source of ROS. Similarly, inflammatory process also induces inducible NO synthase in the pancreas, which gives rise to RNS [11]. In addition to the induction of ROS-/RNS-generating enzymes in acinar cells, recruited inflammatory cells also contribute to the free radical production. Establishment of the acute pancreatitis releases chemical mediators and cytokines from the pancreas, which results in the infiltration of inflammatory cells, such as neutrophil and macrophage. The 'respiratory burst' derived from neutrophils' NADPH-oxidase system produces further oxidative stress in the diseased pancreas [12].

Free radicals also contribute to the systemic inflammatory responses, which cause fatal organ failure. For example, ischemia/reperfusion injury of rat pancreas resulted in lung injury accompanied with increased blood concentrations of nitric oxide and hydroxyl radicals [13]. Administration of the free radical scavenger edaravone showed a protective effect against lung injury in the taurocholate pancreatitis model [14], suggesting the free radical production itself could be a promising therapeutic target in acute pancreatitis. In summary, free radicals in acute pancreatitis mainly affect the disease severity via the activation of the inflammatory cascade, as a promoter of inflammation at the acute phase.

Chronic Pancreatitis and Free Radicals

In contrast to acute pancreatitis, symptoms of chronic pancreatitis are derived from cumulative necrosis-fibrosis sequences. For the establishment of diffuse destruction of exocrine and endocrine pancreas, persistent inflammation is essential. A characteristic feature of the pancreatic fibrosis is the activation of pancreatic stellate cells (PSC) [15], which have a critical role in the inflammation-fibrosis sequences. PSC reside within normal pancreas at a quiescent state, and upon activation by various stimuli including inflammatory cytokines or chemical mediators, start to proliferate and depict the characteristic features of myofibroblasts, such as α-smooth muscle actin expression [16] or extracellular matrix production.

Free radicals are crucial for the activation of PSC. Various inflammatory cytokines have been reported to activate PSC, such as platelet-derived growth factor, interleukin-1β, and angiotensin II. These cytokines activated PSC in vitro, and this activation was uniformly accompanied by increased ROS production within PSC. In vitro activation of PSC, in vivo development of pancreatic fibrosis in WBN/KOB rat and dibutyltin dichloride-induced chronic pancreatitis were attenuated by administration of diphenylene iodonium and apocynin, the inhibitors of NADPH oxidase [17]. These results indicate the indispensable role of the ROS-generating system as an upstream regulator of inflammatory cascade during pancreatic fibrosis.

Another ROS-generating enzyme is also reported to contribute to the PSC activation. Inhibition of superoxide dismutase by diethyldithiocarbamate resulted in the increased lipid peroxidation in cultured PSC, and this treatment increased α-smooth muscle actin-positive cells and type I collagen production, which are characteristic features for activated PSC [18]. Of note, allopurinol, the XOD inhibitor, attenuated this effect, indicating XOD as an additional source of ROS during chronic pancreatitis and PSC activation.

Excessive alcohol consumption and smoking are common risk factors for the development of chronic pancreatitis, and these factors also affect the free radical production, to maintain the persistent inflammation in the pancreas. Alcohol metabolisms are generally divided into two pathways: (a) oxidative pathway which generates acetaldehyde depending on alcohol dehydrogenase and CYP2E1, and (b) nonoxidative pathway which produces fatty acid ethyl esters. These metabolites are reported to injure pancreatic acinar cells, and several pathways involve increased free radical production. Among them, acetaldehyde was reported to activate PSC, with increased lipid peroxidation within PSC [19]. Similarly, chronic exposure to cigarette smoke also caused pancreatic tissue injury in a rat model, which was accompanied by declined activity of glutathione peroxidase in the pancreas [20]. The fact that an exogenous factor such as alcohol intake or smoking could aggravate the fibrotic change indicates the possibility of intervention to overcome the progressive destruction of pancreatic tissue.

Multiple Signaling Pathway Activation by Free Radicals

Free radicals are known to activate multiple signaling pathways in a tissue- and context-specific manner. In acute or chronic pancreatitis, several inflammatory signals play pivotal roles in the establishment of tissue injury. NF-κB was found to activate immunoglobulin κ enhancer in B cells, and was also expressed in pancreatic acinar cells. There are several contradictory results due to the complex regulations and multiple inputs toward this molecule, but cumulative results indicate that ROS activates the NF-κB pathway, especially in acute pancreatitis. The activation of NF-κB leads to the proinflammatory gene expression. The antioxidant *N*-acetylcysteine could attenuate the activation of NF-κB in rat cerulein pancreatitis, suggesting ROS triggers NF-κB activation [21]. In contrast, there are few reports about the contribution of NF-κB in the pathogenesis of chronic pancreatitis, indicating its role as a promoter of acute inflammation.

The other signaling pathway involved in the pancreatic inflammatory change is mitogen-activated protein kinase (MAPK). The MAPK signal is sensitive to the redox status, and participates in numerous inflammatory processes. MAPKs include several classes of molecules, including extracellular-related kinase (ERK), p38MAPK and Jun N-terminal kinase. Recent research has demonstrated that angiotensin II regulates the production of interleukin-6, the inflammatory cytokine, in the acinar cell during the course of acute pancreatitis. This pathway required the activation of MAPK, especially ERK [22], and ERK activation by angiotensin II was attenuated by an antioxidant. This report indicates the essential role of free radicals as signal transduction molecules. Similarly, p38MAPK is also reported to be activated by ROS. Addition of external pressure-activated PSC and p38MAPK activation were also observed along with this change [23]. Epigallocatechin gallate, a potent antioxidant derived from green tea inhibited the activation of p38MAPK in this system. Jun N-terminal kinase was also reported to be activated by hydrogen peroxide and menadione in the isolated rat pancreatic acini [24], together with ERK and p38MAPK. Activation of these MAPKs leads to the enhancement of the inflammatory reaction in acute pancreatitis. In contrast to NF-κB, these MAPK pathways are involved in the pathogenesis of chronic pancreatitis via PSC activation. In addition to these NF-κB and MAPK pathways, activation of the AKT pathway or apoptosis-related pathway by free radicals is also reported. Activation of these multiple pathways regulates cell death, recruitment of inflammatory cells and fibrosis.

Free Radicals as Therapeutic Targets in Pancreatic Inflammation

Conventional therapy such as intravenous fluid administration or antibiotics in acute pancreatitis or pancreatic enzyme supplementation in chronic pancreatitis is a supportive therapy, and mainly focuses on symptom relief or natural recovery. The above-mentioned intriguing roles of free radicals in acute and chronic pancreatitis are now considered to be novel therapeutic targets for the treatment of pancreatitis. Several molecules are reported to scavenge free radicals in vitro and in vivo, and some of them are clinically applicable. Edaravone, the free radical scavenger, which was originally used in the acute-phase stroke patient to prevent the reperfusion injury in the brain, was reported to reduce the ascites and serum amylase level in the closed duodenal loop-induced acute pancreatitis model [25]. Since the edaravone is widely used in the clinical setting already, early application to the acute pancreatitis is expected.

Besides edaravone, several antioxidants were examined as therapeutic agents in acute pancreatitis and chronic pancreatitis. The antioxidants used are glutamine, vitamin C, vitamin E and so on [26]. These antioxidants are reported to reduce the hospitalization in acute pancreatitis patients and to improve the abdominal pain in chronic pancreatitis patients. For a chronic pancreatitis patient, intake of vitamin C or vitamin E is recommended. However, these trials were carried out in a relatively small number of patients, and further validation needs to be carried out in a larger number of patients.

Another approach against the free radicals in pancreatitis is targeting the ROS-/RNS-generating enzymes. XOD is one of the ROS-generating enzymes, and its pharmacological inhibitor already exists. Allopurinol is an XOD inhibitor, which is widely used to prevent the gout attack by reducing uric acid synthesis. Allopurinol also inhibits ROS production, and recent report has suggested the possible therapeutic effect of allopurinol in rat acute pancreatitis model [27]. Another group applied allopurinol as a chemopreventive agent of postendoscopic retrograde cholangiopancreatography (ERCP) pancreatitis; however, this randomized trial concluded that allopurinol does not reduce the overall risk of post-ERCP pancreatitis [28]. According to this study, allopurinol might have some benefits for the high-risk group of post-ERCP

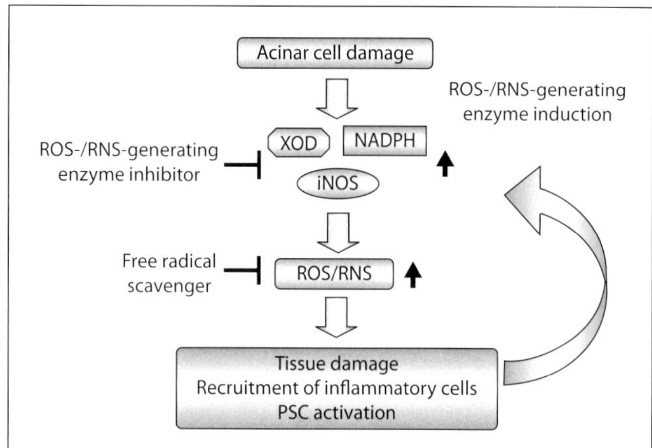

Fig. 1. Schematic view of the positive feedback loop for the establishment of tissue injury by free radicals and the potential therapeutic targets. iNOS = Inducible NO synthase.

pancreatitis. To evaluate the beneficial effect of allopurinol in these patients, further study should be carried out.

Targeting NADPH oxidase could also be a possible antioxidant therapy. Until now, NADPH oxidase inhibition has not been applied in acute or chronic pancreatitis treatment in human. One study demonstrated that apocynin, the NADPH oxidase inhibitor, reduced hydrogen peroxide and nitrate concentrations in exhaled breath [29]. Inhalation of apocynin had no significant adverse effects, and oral administration of apocynin is also possible in animal models. The efficacy of apocynin and other NADPH oxidase inhibitors in acute and chronic pancreatitis should be extensively explored.

Conclusions

Researches have recently revealed the essential roles of free radicals during the establishment of acute pancreatitis and chronic pancreatitis. The involvement of multiple signaling pathway activation by free radicals prompted the development of novel therapies targeting free radicals themselves or free radical-producing systems (fig. 1). Evaluation of novel free radical scavengers and ROS-/RNS-generating enzyme inhibitors will lead to innovative therapies.

References

1 Frossard JL, Steer ML, Pastor CM: Acute pancreatitis. Lancet 2008;371:143–152.

2 Ammann RW, Muellhaupt B: Progression of alcoholic acute to chronic pancreatitis. Gut 1994;35:552–556.

3 Pezzilli R: Chronic pancreatitis: maldigestion, intestinal ecology and intestinal inflammation. World J Gastroenterol 2009;15:1673–1676.
4 Genuth S, Alberti KG, Bennett P, Buse J, Defronzo R, Kahn R, Kitzmiller J, Knowler WC, Lebovitz H, Lernmark A, Nathan D, Palmer J, Rizza R, Saudek C, Shaw J, Steffes M, Stern M, Tuomilehto J, Zimmet P: Follow-up report on the diagnosis of diabetes mellitus. Diabetes Care 2003;26:3160–3167.
5 Raimondi S, Maisonneuve P, Lowenfels AB: Epidemiology of pancreatic cancer: an overview. Nat Rev Gastroenterol Hepatol 2009;6:699–708.
6 Gorelick FS, Thrower E: The acinar cell and early pancreatitis responses. Clin Gastroenterol Hepatol 2009;7:S10–S14.
7 Cuthbertson CM, Christophi C: Disturbances of the microcirculation in acute pancreatitis. Br J Surg 2006;93:518–530.
8 Barlas A, Cevik H, Arbak S, Bangir D, Sener G, Yegen C, Yegen BC: Melatonin protects against pancreaticobiliary inflammation and associated remote organ injury in rats: role of neutrophils. J Pineal Res 2004;37:267–275.
9 Rau B, Poch B, Gansauge F, Bauer A, Nussler AK, Nevalainen T, Schoenberg MH, Beger HG: Pathophysiologic role of oxygen free radicals in acute pancreatitis: initiating event or mediator of tissue damage? Ann Surg 2000;231:352–360.
10 Yu JH, Lim JW, Kim KH, Morio T, Kim H: NADPH oxidase and apoptosis in cerulein-stimulated pancreatic acinar AR42J cells. Free Radic Biol Med 2005;39:590–602.
11 Chvanov M, Petersen OH, Tepikin A: Free radicals and the pancreatic acinar cells: role in physiology and pathology. Philos Trans R Soc Lond B Biol Sci 2005;360:2273–2284.
12 Robinson JM: Phagocytic leukocytes and reactive oxygen species. Histochem Cell Biol 2009;131:465–469.
13 Chen CF, Chen HT, Wang D, Li JP, Fong Y: Restrictive ventilatory insufficiency and lung injury induced by ischemia/reperfusion of the pancreas in rats. Transplant Proc 2008;40:2185–2187.
14 Yang T, Mao YF, Liu SQ, Hou J, Cai ZY, Hu JY, Ni X, Deng XM, Zhu XY: Protective effects of the free radical scavenger edaravone on acute pancreatitis-associated lung injury. Eur J Pharmacol 2010;630:152–157.
15 Masamune A, Watanabe T, Kikuta K, Shimosegawa T: Roles of pancreatic stellate cells in pancreatic inflammation and fibrosis. Clin Gastroenterol Hepatol 2009;7:S48–S54.
16 Masamune A, Shimosegawa T: Signal transduction in pancreatic stellate cells. J Gastroenterol 2009;44:249–260.
17 Masamune A, Watanabe T, Kikuta K, Satoh K, Shimosegawa T: NADPH oxidase plays a crucial role in the activation of pancreatic stellate cells. Am J Physiol Gastrointest Liver Physiol 2008;294:G99–G108.
18 Tanioka H, Mizushima T, Shirahige A, Matsushita K, Ochi K, Ichimura M, Matsumura N, Shinji T, Tanimoto M, Koide N: Xanthine oxidase-derived free radicals directly activate rat pancreatic stellate cells. J Gastroenterol Hepatol 2006;21:537–544.
19 Apte MV, Phillips PA, Fahmy RG, Darby SJ, Rodgers SC, McCaughan GW, Korsten MA, Pirola RC, Naidoo D, Wilson JS: Does alcohol directly stimulate pancreatic fibrogenesis? Studies with rat pancreatic stellate cells. Gastroenterology 2000;118:780–794.
20 Jianyu-Hao, Guang-Li, Baosen-pang: Evidence for cigarette smoke-induced oxidative stress in the rat pancreas. Inhal Toxicol 2009;21:1007–1012.
21 Gukovsky I, Gukovskaya AS, Blinman TA, Zaninovic V, Pandol SJ: Early NF-kappaB activation is associated with hormone-induced pancreatitis. Am J Physiol 1998;275:G1402–G1414.
22 Chan YC, Leung PS: Involvement of redox-sensitive extracellular-regulated kinases in angiotensin II-induced interleukin-6 expression in pancreatic acinar cells. J Pharmacol Exp Ther 2009;329:450–458.
23 Asaumi H, Watanabe S, Taguchi M, Tashiro M, Otsuki M: Externally applied pressure activates pancreatic stellate cells through the generation of intracellular reactive oxygen species. Am J Physiol Gastrointest Liver Physiol 2007;293:G972-G978.
24 Dabrowski A, Boguslowicz C, Dabrowska M, Tribillo I, Gabryelewicz A: Reactive oxygen species activate mitogen-activated protein kinases in pancreatic acinar cells. Pancreas 2000;21:376–384.
25 Araki Y, Andoh A, Yokono T, Asano N, Yoshikawa K, Bamba S, Ishizuka I, Fujiyama Y: The free radical scavenger edaravone suppresses experimental closed duodenal loop-induced acute pancreatitis in rats. Int J Mol Med 2003;12:121–124.
26 Mohseni Salehi Monfared SS, Vahidi H, Abdolghaffari AH, Nikfar S, Abdollahi M: Antioxidant therapy in the management of acute, chronic and post-ERCP pancreatitis: a systematic review. World J Gastroenterol 2009;15:4481–4490.
27 Comert B, Isik AT, Aydin S, Bozoglu E, Unal B, Deveci S, Mas N, Cinar E, Mas MR: Combination of allopurinol and hyperbaric oxygen therapy: a new treatment in experimental acute necrotizing pancreatitis? World J Gastroenterol 2007;13:6203–6207.

28 Romagnuolo J, Hilsden R, Sandha GS, Cole M, Bass S, May G, Love J, Bain VG, McKaigney J, Fedorak RN: Allopurinol to prevent pancreatitis after endoscopic retrograde cholangiopancreatography: a randomized placebo-controlled trial. Clin Gastroenterol Hepatol 2008;6:465–471.

29 Stefanska J, Sokolowska M, Sarniak A, Wlodarczyk A, Doniec Z, Nowak D, Pawliczak R: Apocynin decreases hydrogen peroxide and nitrate concentrations in exhaled breath in healthy subjects. Pulm Pharmacol Ther 2010;23:48–54.

Tooru Shimosegawa
1-1, Seiryo-machi, Aobaku
Sendai, Miyagi, 980-8574 (Japan)
Tel. +81 22 717 7171, Fax +81 22 717 7177, E-Mail tshimosegawa@int3.med.tohoku.ac.jp

Naito Y, Suematsu M, Yoshikawa T (eds): Free Radical Biology in Digestive Diseases.
Front Gastrointest Res. Basel, Karger, 2011, vol 29, pp 164–171

Free Radicals and Cancer Treatment

Satoshi Kokura

Department of Molecular Gastroenterology and Hepatology, Kyoto Prefectural University of Medicine, Kyoto, Japan

Abstract

It is thought that the free radical response is involved in the process of chemical carcinogenesis. Various kinds of carcinogens `become free radicals in vivo, and cancer-causing initiation occurs by binding to the DNA of free radicals. Also, the generation of the reactive oxygen species (ROS) is said to have a cancer-causing potential. On the other hand, many cancer therapies now use ROS and free radical reaction. NF-κB, which is a transcription factor associated with cancer metastasis and tumor growth, is an important target of cancer therapy. However, ROS are involved in the activation of this NF-κB. Therefore, the production of ROS in the tumor becomes a double-edged sword by the degree. This report deals with cancer therapy from the perspective of free radicals.

Anoxia/Reoxygenation and Metastasis of Cancer

Metastases of cancers are assumed to occur in multiple steps. The initial step is the detachment of cells from primary tumors. The integrity and morphology of cancer cell-cell adhesions are maintained by E-cadherin and its associated intracellular catenin molecules. In vitro experimental evidence has suggested that selective loss of E-cadherin causes disruption of cell-cell adhesion [1] and acquisition of invasiveness [2]. Furthermore, immunohistochemical studies have indicated that decreased E-cadherin expression in various human cancers in vivo is correlated with invasive and metastatic potential [3]. These findings indicate that downregulation of E-cadherin facilitates tumor cell invasion by causing separation of cells from the primary tumor. However, mutational inactivation of E-cadherin appears to be common only in diffuse-type gastric carcinomas and infiltrating lobular breast carcinomas [4]. In the majority of cancers with altered E-cadherin expression, such as colorectal cancers and esophageal cancers, gene mutations are rare or absent [5]. Thus, researchers are interested in understanding the mechanisms by which the expression of E-cadherin is reduced or eliminated. Cancer tissues develop a pathophysiological microenvironment during growth, characterized by an irregular microvascular network and regions of chronically and transiently ischemic cells. Ischemic cancer cells are known to be reperfused by neovascularization or by a decrease in tissue pressure. Bush et al. [6] reported that 3 h of ischemia clearly reduced the total amount of E-cadherin in canine kidney cells. Rofstad and Halsor [7] indicated that tumor hypoxia promotes the development of metastasis. These findings suggest that anoxia/reoxygenation induces the reduction of E-cadherin expression, which may promote metastasis.

Our previous study provided evidence to support the hypothesis that anoxia/reoxygenation induces the reduction of E-cadherin expression in human colon adenocarcinoma cell lines [8]. The reduction of E-cadherin on cancer cell surfaces causes the release of cancer cells from the primary tumors [9], which is the initial step of cancer metastasis. Our results showing that E-cadherin on the plasma membrane had almost disappeared and the attachments of cancer cells to cancer cells were partially destroyed after 22 h of reoxygenation suggest that anoxia/reoxygenation may induce cancer metastasis. Anoxia can induce the expression of several adhesion molecules on epithelial cells, endothelial cells, and cancer cells. Many studies have indicated that the expression of adhesion molecules is important for tumor invasion and metastasis in colorectal cancer [10]. The loss of E-cadherin expression might be related to the invasive capacity as well as metastatic potential of cancer cells [11]. Previous studies have demonstrated that downregulation of E-cadherin expression is associated with differentiation grade and metastasis in human colorectal carcinomas [12] and squamous cell carcinomas of the head and neck [13]. Expression of E-cadherin has been studied in tissue sections from a variety of differentiation grades, and lower levels of expression were observed in more poorly differentiated cancers [14]. The downregulation of E-cadherin is not necessarily a causative factor of metastasis to the liver. For example, differentiated gastric cancers, which express high amounts of E-cadherin, often metastasize to the liver [15]. In addition, well-differentiated colorectal carcinomas frequently develop liver metastasis. However, at the initial step of metastasis, it is necessary for cancer cells to detach from the primary tumor. Furthermore, the downregulation of E-cadherin expression may be essential for cancer cells to detach from the primary tumor [9]. Thus, the transient downregulation of E-cadherin, by which cancer cells acquire their invasive property, must be organized by some mechanism at the primary site in the process of liver metastasis. The expression of E-cadherin was significantly reduced during reoxygenation from 4 to 22 h. Western blots of colon carcinoma cells exposed to normoxia or anoxia revealed two different patterns of downregulation, depending on the duration of reoxygenation. Four hours of reoxygenation downregulated the expression of E-cadherin, demonstrated by immunofluorescent staining and enzyme-linked immunosorbent assays. However, no loss in the total amount of E-cadherin was detectable by Western blotting after 4 h of reoxygenation, which indicates that the reduction in E-cadherin expression is probably the result of the internalization of surface-accessible E-cadherin. Conversely, Western blots revealed a clear reduction in the total amount of E-cadherin after 22 h of reoxygenation, suggesting that the loss of E-cadherin is not only the result of the internalization; however, this additional mechanism is unclear. Nakayama et al. [16] showed that E-cadherin expression was decreased in highly metastatic cell lines, whereas the E-cadherin mRNA levels did not differ between highly metastatic cell lines and weakly metastatic cell lines [17]. Taken together with another previous report [6], it is suggested that E-cadherin on the cell surface is internalized 4 h after reoxygenation, and degraded in response to further reoxygenation (22 h).

The anoxia/reoxygenation-induced reduction of E-cadherin expression is transient. The expression of E-cadherin and the cancer cell-cell adhesion were recovered 46 h after reoxygenation. Osada et al. [18] reported that E-cadherin participates in secondary tumor formation in the liver. They speculated that increased cell-cell adhesion causes self-aggregation of cancer cells and promotes tumor formation in the liver. This scenario is supported by a previous report by Mayer and colleagues, which shows by immunohistochemical analysis that all liver metastases of gastric cancers showed a high frequency of strongly E-cadherin-positive cells, regardless of the staining pattern in the primary cancer [19]. To initiate metastasis, cancer cells have to become detached from the primary cancer, and in this step downregulation of E-cadherin is known to play an important role. However, primary colorectal cancers and their liver metastatic nodules often show a well-differentiated histology and retain their E-cadherin expression. Differentiated human gastric cancer, which retains E-cadherin expression, often metastasizes to the liver, whereas undifferentiated gastric cancers, which show reduced amounts of E-cadherin, often show intraperitoneal dissemination and lymphogenous metastasis but not liver metastasis. These findings suggest that in the process of liver metastasis, the transient downregulation of E-cadherin, by which cancer cells acquire their invasive property, is organized by some mechanism at the primary site. In the liver, E-cadherin, which is downregulated at the primary site, might become recovered and thus play an important role in secondary cancer formation.

NF-κB as a Molecular Target of Cancer

NF-κB is generally believed to be activated by various external stresses, cytokines and drugs. The common thread in NF-κB activation is the intracellular production of reactive oxygen species (ROS) [19–21]. Therefore, ROS are considered the common pathway for NF-κB activation. In fact, free radical scavengers suppress NF-κB activation by various stimuli.

Recent studies have demonstrated that transcription factor NF-κB is constitutively activated in various cancer tissues [22–24]. Previous studies have shown that NF-κB regulates the expression of multiple genes, including angiogenic factors and anti-apoptotic protein. Inhibition of NF-κB activity has thus been proposed to suppress tumor growth and cancer metastasis by inhibiting angiogenesis, and by increasing

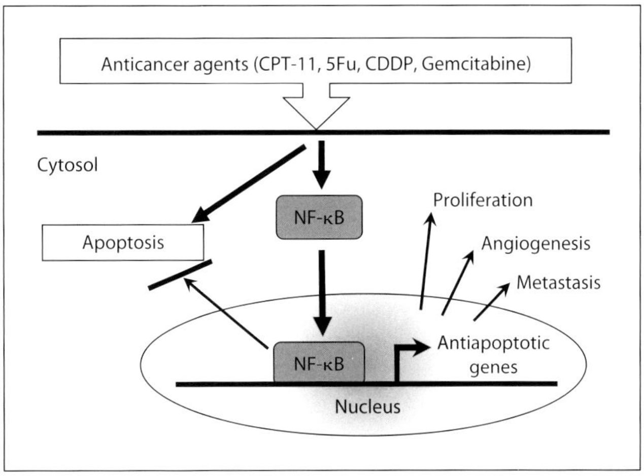

Fig. 1. Anticancer drug causes cancer cell apoptosis. However, anticancer drug makes NFκB activate through ROS. As a result, the apoptotic effects of anticancer drug in itself are weakened, and the survival signals such as proliferation, metastasis or the neovascularization of the cancer cell work at the same time.

apoptosis. Several studies have shown that NF-κB plays an important antiapoptotic role. A previous study in our laboratory showed that inhibition of NF-κB in a human gastric cancer cell line resulted in the potentiation of apoptosis in response to TNF-α.

Inhibition of NF-κB by Radical Scavenger Enhances the Anticancer Effect of Cytotoxic Agent

When cancer cells are exposed to CPT-11 (anti-cancer agent: topoisomelase-1 inhibitor), the transcription factor NF-κB is activated [25, 26]. Several studies have reported that NF-κB activation, induced by exposing cancer cells to TNF-α, CPT-11 and other anticancer agents, has an antiapoptotic effect on cancer cells (fig. 1). While the mechanism of these antiapoptotic effects has not been fully elucidated, antiapoptotic proteins that are expressed as a result of NF-κB activation are believed to be involved. Our previous study showed that when colon26 cancer cells were exposed to CPT-11, NF-κB in colon26 cancer cells was activated in a dose-dependent manner. In addition, CPT-11 was cytotoxic to colon26 cells in a dose-dependent manner, and the cytotoxic effects of CPT-11 were enhanced in the presence of an inhibitor of NF-κB activation. In other words, in colon26 cells, CPT-11 induces apoptosis and suppresses apoptosis via NF-κB, and consequently, cytotoxicity can be augmented by blocking the latter action.

How then is NF-κB activated when colon26 cells are exposed to SN38? NF-κB can be activated by various external stresses, cytokines and drugs. The common thread in NF-κB induction by these stimuli is the intracellular production of ROS. Therefore, ROS are believed to be the common pathway for NF-κB activation. We thus investigated the production of ROS in colon26 cancer cells that were exposed to CPT-11. The results showed that CPT-11 induced the production of ROS in a dose-dependent

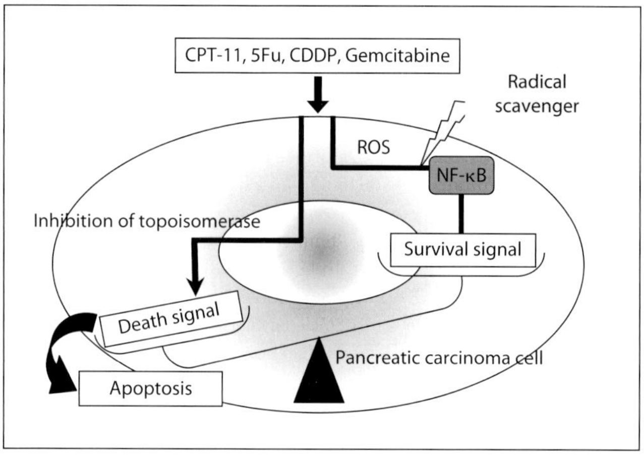

Fig. 2. When activation of NF-κB is caused by ROS, which is produced by anticancer drug, the activation of NF-κB can be inhibited using a radical scavenger. As a result, only the death signal of the anticancer drug acts.

manner. It was then necessary to determine whether ROS were involved in NF-κB activation. We showed that edaravone eliminated the ROS produced in the cancer cells by CPT-11 in a dose-dependent manner and suppressed NF-κB activation [27]. While it has already been reported that suppression of NF-κB activation enhances the cytotoxicity of TNF-α and CPT-11, the results of our study also showed that combination therapy of edaravone and CPT-11 augmented the cytotoxicity of CPT-11. Moreover, it was confirmed that this augmentation in cytotoxicity was attributable to increased apoptosis (fig. 2). How then does edaravone-induced suppression of NF-κB augment CPT-11-induced apoptosis? For TNF-α to induce apoptosis via the caspase cascade, inhibition of protein synthesis is required, and activation of NF-κB due to TNF-α appears to facilitate the synthesis of a protein blocking the caspase cascade. It has been reported that when NF-κB is activated by TNF-α, antiapoptotic proteins such as IAP-1 and IAP-2 [28], are synthesized, and these proteins block the caspase cascade. Like TNF-α, anticancer agents also induce apoptosis by activating the caspase cascade. IAP-1 and IAP-2 expression can suppress etoposide-induced apoptosis, which is consistent with the ability of these proteins to block activation of caspase-3. The level of caspase-3 activation for the combination therapy of CPT-11 and edaravone was greater than that of CPT-11 monotherapy. This may have caused an increase in apoptotic cells, but to the best of our knowledge, expression of antiapoptotic proteins is totally unaffected by CPT-11 monotherapy or combination therapy of CPT-11 and edaravone. Therefore, the mechanism of caspase-3 activation associated with combination therapy of CPT-11 and edaravone has not been clarified. With regard to the mechanisms for caspase cascade activation and apoptosis induced by anticancer agents, several studies have documented that upregulation of the Fas ligand is involved in several types of cancer cells. It is thus possible that NF-κB activation suppresses Fas-mediated cell killing.

Preclinical studies are being conducted regarding the inhibition of NF-κB activation in several laboratories, including ours. Wang et al. [29] reported one of the most effective methods for inhibiting NF-κB activity based on recombinant adenovirus-mediated overexpression of the IkBα gene. However, the clinical feasibility of gene therapy for inhibiting NF-κB activity is limited by the intratumoral delivery of a recombinant virus expressing an NF-κB inhibitor. In actual anticancer treatments, therapy for widely disseminated metastatic diseases is more important than localized therapy. Therefore, systemic therapeutic agents that block the antiapoptotic response mediated by chemotherapy-induced NF-κB activation in cancer cells could improve the efficacy of chemotherapy in systemic cancer. Edaravone is already being used as a therapeutic drug for treatment of cerebrovascular diseases. Its free radical-scavenging effects and safety in humans have been established. With regard to the effects of combination therapy of CPT-11 and edaravone on our subcutaneous tumor model, the results suggest that edaravone suppressed the activation of NF-κB induced by CPT-11, thus enhancing apoptosis and inhibiting tumor growth. Furthermore, with our lung metastasis model, although the number of metastatic nodules for combination therapy of edaravone and CPT-11 was significantly lower, there was no marked difference in the size of nodules. This suggests that the combination therapy of CPT-11 and edaravone not only enhances apoptosis, but also is involved in the early stages of pulmonary metastasis formation, such as colonization, infiltration and angiogenesis. This is supported by studies documenting that Cu-ZnSOD and EC-SOD suppress pulmonary metastases [29]. Although a variety of mechanisms, including inhibition of NF-κB activation, may potentially contribute to the efficacy of the combination therapy of edaravone and CPT-11, additional studies are necessary.

Hyperthermia Enhances Free Radical-Dependent Cytotoxicity of Polyunsaturated Fatty Acids

Previous studies have indicated that polyunsaturated fatty acids can induce the suppression of cancer growth and that this action is free radical dependent. Many polyunsaturated fatty acids including γ-linolenic acid (GLA) induce superoxide generation by cancer cells, and the generation of superoxide is the main mechanism for the anticancer action of polyunsaturated fatty acids. Neutrophils are known to generate superoxide in a temperature-dependent manner in response to several stimuli. Our previous study indicated that neutrophils most strongly generate superoxide at 41°C. On the other hand, in a cell-free system, DNA produces superoxide by stimulus of cisplatin, and hyperthermia enhances this production. Hyperthermia also enhances the generation of superoxide from rat hepatocellular carcinoma upon stimulation by polyunsaturated fatty acids, in particular GLA.

Our previous studies have demonstrated that the combination of GLA with hyperthermia enhances the antitumor effect of hyperthermia on rat hepatocellular carcinoma in vivo [30]. These previous studies also showed that the enhanced antitumor effect of the combination of hyperthermia and GLA is due to the increase in free radical reactions in tumor tissue [30]. Hyperthermia enhances the cytotoxicity of GLA and that superoxide dismutase (SOD) or α-tocopherol significantly attenuates the cytotoxicity of GLA suggests that hyperthermic enhancement of superoxide generation from carcinoma cells may contribute to their cytotoxicity. A positive correlation between the ability of fatty acids to generate superoxide and cytotoxicity of fatty acids was shown. Although the efficacy of combining polyunsaturated fatty acids and radiation, chemotherapy, or hyperthermia is relatively well known, the details of the mechanisms by which the cytotoxicity is enhanced remain incomplete. We indicate that the transferred GLA itself is responsible for the cell toxicity, although whether the GLA is taken up into the cell or remains on the membrane surface is unclear. The coexistence of α-tocopherol inhibits GLA-induced cytotoxicity in various cells such as HeLa and leukemic cells. Interestingly, exogenous addition of SOD or SOD plus catalase inhibits GLA combined with hyperthermia-induced cytotoxicity. Thus, superoxide generated on the outside of the cells appears to be responsible for cytotoxicity. There have been reports that oxygen radicals contribute to apoptosis in various cells. Activation of superoxide generation in neutrophils has been suggested to be associated with the influx of Ca^{2+} into the cytoplasm from extracellular space. Furthermore, polyunsaturated fatty acids have been shown to increase the intracellular concentration of Ca^{2+}. Recently, Kameda et al. [31] reported that hyperthermia increases the intracellular concentration of Ca^{2+} in the U937 cell line and that the intracellular Ca^{2+} increase plays a crucial role in apoptosis induced by hyperthermia. Taken together with these reports, we suggest that GLA plus hyperthermia activates superoxide generation by carcinoma cells due to an increase in the intracellular Ca^{2+} level.

References

1 Takeichi M: Cadherins: A molecular family important in selective cell-cell adhesion. Annu Rev Biochem 1990;59:237–252.

2 Behrens J, Mareel MM, Van Roy FM, Birchmeier WJ: Dissecting tumor cell invasion: epithelial cells acquire invasive properties after the loss of uvomorulin-mediated cell-cell adhesion. Cell Biol 1989; 108:2435–2447.

3 Shimoyama Y, Hirohashi S: Expression of E- and P-cadherin in gastric carcinomas. Cancer Res 1991; 51:2185–2192.

4 Berx G, Becker KF, Hofler H, Van Roy F: Mutations of the human E-cadherin (CHD1) gene. Hum Mutat 1998;12:226–237.

5 Ilyas M, Tomlinson IP, Hanby A, Talbot IC, Bodmer WF: Allele loss, replication errors and loss of expression of E-cadherin in colorectal cancers. Gut 1997; 40:654–659.

6 Bush KT, Tsukamoto T, Nigam SK: Selective degradation of E-cadherin and dissolution of E-cadherin-catenin complexes in epithelial ischemia. Am J Physiol Renal Physiol 2000;278:F847–F852.

7 Rofstad EK, Halsor EF: Hypoxia-associated spontaneous pulmonary metastasis in human melanoma xenografts: involvement of microvascular hot spots induced in hypoxic foci by interleukin 8. Br J Cancer 2002;86:301–308.

8 Kokura S, Yoshida N, Imamoto E, Ueda M, Ishikawa T, Uchiyama K, Kuchide M, Naito Y, Okanoue T, Yoshikawa T: Anoxia/reoxygenation down-regulates the expression of E-cadherin in human colon cancer cell lines. Cancer Lett 2004;211:79–87.

9 Birchmeier W, Weidner KM, Hulsken J, Behrens J: Molecular mechanisms leading to cell junction (cadherin) deficiency in invasive carcinomas. Semin Cancer Biol 1993;4:231–239.

10 Hoff SD, Irimura T, Matsushita Y, Ota DM, Cleary KR, Hakomori S: Metastatic potential of colon carcinoma. Expression of ABO/Lewis-related antigens. Arch Surg 1990;125:206–209.
11 Nigam AK, Savage FJ, Boulos PB, Stamp GW, Liu D, Pignatelli M: Loss of cell-cell and cell-matrix adhesion molecules in colorectal cancer. Br J Cancer 1993;68:507–514.
12 van der Wurff AA, Arends JW, van der Linden EP, ten Kate J, Bosman FT: L-CAM expression in lymph node and liver metastases of colorectal carcinomas. J Pathol 1994;172:177–181.
13 Schipper JH, Frixen UH, Behrens J, Unger A, Jahnke K, Birchmeier W: E-cadherin expression in squamous cell carcinomas of head and neck: inverse correlation with tumor dedifferentiation and lymph node metastasis. Cancer Res 1991;51:6328–6337.
14 Shimoyama Y, Hirohashi S: Cadherin intercellular adhesion molecule in hepatocellular carcinomas: loss of E-cadherin expression in an undifferentiated carcinoma. Cancer Lett 1991;57:131–135.
15 Esaki Y, Hirayama R, Hirokawa K: A comparison of patterns of metastasis in gastric cancer by histologic type and age. Cancer 1990;65:2086–2090.
16 Nakayama Y, Okazaki K, Shibao K, Sako T, Hirata K, Nagata N, Kuwano M, Itoh H: Alterative expression of the collagenase and adhesion molecules in the highly metastatic clones of human colonic cancer cell lines. Clin Exp Metastasis 1998;16:461–469.
17 Kitadai Y, Bucana CD, Ellis LM, Anzai H, Tahara E, Fidler IJ: In situ mRNA hybridization technique for analysis of metastasis-related genes in human colon carcinoma cells. Am J Pathol 1995;147:1238–1247.
18 Osada T, Sakamoto M, Ino Y, Iwamatsu A, Matsuno Y, Muto T, Hirohashi S: E-cadherin is involved in the intrahepatic metastasis of hepatocellular carcinoma. Hepatology 1996;24:1460–1467.
19 Ichikawa H, Flores S, Kvietys PR, Wolf RE, Yoshikawa T, Granger DN, Aw TY: Molecular mechanisms of anoxia/reoxygenation-induced neutrophil adherence to cultured endothelial cells. Circ Ress 1997;81:922–931.
20 Lo SK, Janakidevi K, Lai L, Malik AB: Hydrogen peroxide-induced increase in endothelial adhesiveness is dependent on ICAM-1 activation. Am J Physiol 1993;264:L406–L412.
21 Osborn L: Leukocyte adhesion to endothelium in inflammation. Cell 1990;623–626.
22 Sasaki N, Morisaki T, Hashizume K, Yao T, Tsuneyoshi M, Noshiro H, Nakamura K, Yamanaka T, Uchiyama A, Tanaka M, Katano M: Nuclear factor-κB p65 (RelA) transcription factor is constitutively activated in human gastric carcinoma tissue. Clin Cancer Res 2001;7:4136–4142.
23 Wang W, Abbruzzese JL, Evans DB, Larry L, Cleary KR, Chiao PJ: The nuclear factor-κB RelA transcription factor is constitutively activated in human pancreatic adenocarcinoma cells. Clin Cancer Res 1999; 119–127.
24 Sovak MA, Bellas RE, Kim DW, Zanieski GJ, Rogers AE, Traish AM, Sonenshein GE: Aberrant nuclear factor-κB/Rel expression and the pathogenesis of breast cancer. J Clint Invest 1997;100:2952–2960.
25 Piret B, Piette J: Topoisomerase poisons activate the transcription factor NF-kappaB in ACH-2 and CEM cells. Nucleic Acids Res 1996;24:4242–4248.
26 Wang CY, Mayo MW, Baldwin AS Jr: TNF- and cancer therapy-induced apoptosis: potentiation by inhibition of NF-kappaB. Science 1996;274:784–787.
27 Kokura S, Yoshida N, Sakamoto N, Ishikawa T, Takagi T, Higashihara H, Nakabe N, Handa O, Naito Y, Yoshikawa T: The radical scavenger edaravone enhances the anti-tumor effects of CPT-11 in murine colon cancer by increasing apoptosis via inhibition of NF-kappaB. Cancer Lett 2005;229:223–233.
28 Roy N, Deveraux QL, Takahashi R, Salvesen GS, Reed JC: The c-IAP-1 and c-IAP-2 proteins are direct inhibitors of specific caspases. EMBO J 1997; 16:6914–6925.
29 Wang CY, Cusack JC Jr, Liu R, Baldwin AS Jr: Control of inducible chemoresistance: enhanced anti-tumor therapy through increased apoptosis by inhibition of NF-kappaB. Nat Med 1999;5:412–417.
30 Kokura S, Yoshikawa T, Kaneko T, Iimuma S, Nishimura S, Matsuyama K, Naito Y, Yoshida N, Kondo M: Efficacy of hyperthermia and polyunsaturated fatty acids on experimental carcinoma. Cancer Res 1997;57:2200–2202.
31 Kameda K, Kondo T, Tanabe K, Zhao QL, Seto H: The role of intracellular Ca^{2+} in apoptosis induced by hyperthermia and its enhancement by verapamil in U937 cells. Int J Radiat Oncol Biol Phys 2001;49: 1369–1379.

Satoshi Kokura
Department of Molecular Gastroenterology and Hepatology, Kyoto Prefectural University of Medicine
Kawaramachi - Hirokaji, Kyoto 602-8566 (Japan)
Tel. +81 75 251 5519, Fax +81 75 251 0710, E-Mail s-kokura@koto.kpu-m.ac.jp

Author Index

Subject Index